Safety Integrity Level Selection

Systematic Methods Including Layer of Protection Analysis

Safety Integrity Level Selection

Systematic Methods Including Layer of Protection Analysis

Edward M. Marszal, P.E., C.F.S.E.
Dr. Eric W. Scharpf, MIPENZ

Notice

The information presented in this publication is for the general education of the reader. Because neither the author nor the publisher has any control over the use of the information by the reader, both the author and the publisher disclaim any and all liability of any kind arising out of such use. The reader is expected to exercise sound professional judgment in using any of the information presented in a particular application.

Additionally, neither the author nor the publisher has investigated or considered the effect of any patents on the ability of the reader to use any of the information in a particular application. The reader is responsible for reviewing any possible patents that may affect any particular use of the information presented.

Any references to commercial products in the work are cited as examples only. Neither the author nor the publisher endorses any referenced commercial product. Any trademarks or tradenames referenced belong to the respective owner of the mark or name. Neither the author nor the publisher makes any representation regarding the availability of any referenced commercial product at any time. The manufacturer's instructions on the use of any commercial product must be followed at all times, even if in conflict with the information in this publication.

Printed in the United States of America.
10 9 8 7 6 5 4 3

ISBN: 978-1-945541-50-6

ISA
67 T. W. Alexander Drive
P.O. Box 12277
Research Triangle Park, NC 27709

Library of Congress Cataloging-in-Publication Data

Marszal, Edward M.
Safety integrity level selection: systematic methods including layer of protection analysis / Edward M. Marszal, Eric W. Scharpf.
p. cm.
Includes bibliographical references and index.
ISBN 978-1-945541-50-6
1. Industrial safety--Data processing. 2. System safety. I. Scharpf, Eric William. II. Title.
T55 .M3563 2002
620.8'6--dc21

2002003324

Dedication

EMM: For all of my girls: Lisa, Melanie Jane, and Lucy.

EWS: For Susan.

I've done the math
Enough to know
The dangers of
Our second guessing.

— *Tool ("Schism," Lateralus, 2001)*

They do not preach that their God will rouse them a little before the nuts work loose

They do not teach that His pity allows them to drop their job when they dam'-well choose

As in the thronged and lighted ways, so in the dark and desert they stand,

Wary and watchful all their days that their brethren's days may be long in this land.

— *Rudyard Kipling ("The Sons of Martha," 1907)*

Karma police, arrest this man,
He talks in Math.

— *Radiohead ("Karma Police," OK Computer, 1997)*

Contents

Preface

This book describes a systematic method for selecting safety integrity levels (SILs) for safety instrumented systems (SIS). Although numerous methods have been proposed and adopted by industry, layer of protection analysis (LOPA) is rapidly becoming the most frequently used method. Its popularity stems from its ease of use and the accuracy of the results it provides. This LOPA method, more than any other, accounts for most existing layers of protection. With this proper accounting, the SIS is neither overdesigned nor overpriced. The LOPA method ensures that users achieve the maximum return on their risk reduction investments.

We wrote this book because we found that there is a need for a comprehensive and authoritative discussion of the process of selecting SILs. The small amount of literature on the subject is scattered among various periodicals and symposia. Moreover, much of this material is of marginal quality, mainly focusing on qualitative methods.

The result of using poor methods to select SILs is typically either an overdesigned or an underdesigned safety instrumented system. The risk analysis that forms the basis for SIL selection, however, can be greatly improved. This will provide the user with more accurate results so formerly inflated requirements can be relaxed, which will in turn lower not only the initial installation costs, but the cost of ongoing maintenance. Because of the high costs associated with poor selection methods, many practitioners are turning to more quantitative methods, one of which is layer of protection analysis. Thus, layer of protection analysis already boasts a strong and rapidly growing base of sophisticated users.

In developing the tools and procedures that control systems engineers can use to select SILs, we found there was no need for new scientific theories or extensive laboratory research. Instead, these tools and procedures are directly derived from the specific application of general principles of loss prevention engineering to SIS design. The key purpose of this book is to make this sometimes obscure theory accessible to a wider audience and to focus these principles on the task of SIL selection. We are indebted to the late Frank P. Lees for making this task manageable. His three-volume collection, *Loss Prevention for the Process Industries* (1992), contains a vast and vital storehouse of knowledge on the topic of loss prevention.

About this Book

The material for this book was developed from a series of training courses and seminars we have written and delivered over the past few years. The Exida training course (which bears the same name as this book and is co-sponsored by ISA under the catalog number EX-01) and numerous related on-line courses provided a major source of material for this book and also provided an outline for organizing its contents.

Much of the material in this book is based on the application of the safety life cycle as it is described in the international standards ANSI/ISA-84.01-1996-Application of Safety Instrumented Systems for the Process Industry and IEC 61508/61511. This book expands upon the framework developed in these standards. In addition to describing the tasks that users should perform during the safety life cycle, this book also provides detailed procedures for accomplishing these tasks. These procedures are based on risk analysis and reliability engineering principles from a variety of disciplines.

This book is intended to demonstrate the application of quantitative risk analysis techniques and tools to the problem of selecting SILs. Its goal is to bring this topic down to earth and explain it in a clear and approachable way, distilling the essential theory into a format that the practicing control systems engineer can apply quickly and effectively in everyday work. This book is not intended to be a generic theoretical dissertation, nor a comprehensive treatment of the topic of quantitative risk analysis. It presents a focused process for applying simple, yet powerful, tools of quantitative risk analysis specifically to the problem of selecting SILs for safety instrumented systems.

About the Authors

Edward M. Marszal, P.E., C.F.S.E., and Dr. Eric W. Scharpf, MIPENZ, are principal engineers and partners in Exida, an engineering consulting firm that helps users and vendors of automation systems develop safety-critical and high-availability automation solutions. At Exida, both authors are responsible for safety life cycle services for end users, including process hazards analysis, SIL selection and verification, and functional safety assessment of safety critical systems.

Mr. Marszal started his career with UOP, a licensor of process units to the petroleum and petrochemical industries, where he performed functional assessments of control and safety instrumented systems at customer sites worldwide. At UOP, he designed and managed the development of custom control and SIS projects. After leaving UOP, he joined the Environmental Resource Management companies in their Business Risk Solutions consulting group. In this position, he specialized in financial risk analysis and process safety management. He performed and managed risk assessment projects that involved quantitative risk analysis, including preparing Environmental Protection Agency (EPA) Risk Management Plans with off-site consequence analysis for over one hundred facilities. Companies used his recommendations from these projects to ensure regulatory compliance, justify risk reduction expenditures, and optimize insurance coverage.

Mr. Marszal holds a Bachelor of Science in chemical engineering from Ohio State University and is a registered professional engineer in the states of Illinois and Ohio. He has developed and taught safety instrumented system engineering courses for ISA, for whose local chapters in Columbus, Ohio, he holds several executive positions. He is also a member of the American Institute of Chemical Engineers. Mr. Marszal was among the first group of engineers to be awarded the status of Certified Functional Safety Expert (C.F.S.E.) by TÜV Product Services.

Dr. Scharpf has worked as a process chemical engineer in the petroleum and chemicals industries for both Mobil and Air Products and Chemicals in the United States and Europe. In these roles he has designed and developed several new processes and published numerous patents and papers on his work. He has focused much of his career on process optimization, new process design, safety and risk analysis in various segments of the chemical processing industry. This work has included hazard, risk, and consequence analysis as well as safety system work. Because of Dr. Scharpf's increasing responsibility level and personal interest in the safety and risk-related aspects of these systems and processes, in 2000 he joined Mr. Marszal in forming Exida to pursue this work more directly. At Exida, he now leads the consulting, training, and support for safety-critical and high-availability process automation in the Asia-Pacific region. In this role, he has authored and reviewed numerous Exida safety training courses and related material focusing primarily on IEC 61508, 61511, and 62061 safety life cycle applications.

Dr. Scharpf has a Bachelor of Science in chemical engineering from the University of Delaware and a Ph.D. in chemical engineering from Princeton University. Dr. Scharpf is a registered engineer and member of the Institution of Professional Engineers New Zealand and is a member of the New Zealand Society for Risk Management. He also serves as a member of the Board of Directors of the Certified Functional Safety Expert Governance Board. Dr. Scharpf is currently based near Dunedin, New Zealand, and teaches courses in safety, process engineering, and related subjects at the University of Otago in addition to his responsibilities at Exida.

CHAPTER 1

Selecting Safety Integrity Levels: Introduction

The purpose of a *safety instrumented system* (SIS) is to reduce the risk that a process may become hazardous to a tolerable level. The SIS does this by decreasing the frequency of unwanted accidents. The amount of risk reduction that an SIS can provide is represented by its *safety integrity level* (SIL), which is defined as a range of probability of failure on demand. An SIS senses hazardous conditions and then takes action to move the process to a safe state, preventing an unwanted accident from occurring.

The method organizations use to select SILs should be based on their risk of accident, an evaluation of the potential consequences and likelihoods of an accident, and an evaluation of the effectiveness of all relevant process safeguards. Implementing an SIS, and therefore selecting an SIL, should involve considering relevant laws, regulations, and national and international standards. In the United States, the "Process Safety Management" (PSM) section of the OSHA standard OSHA 29 CFR Part 1910.119 requires organizations to provide assurance of the mechanical integrity of all their emergency shutdown systems and safety critical controls. The "Seveso Directive" (96/82/EC) promulgates similar requirements in the European Union. In the United States, ISA—The Instrumentation, Systems, and Automation Society promulgated industry standard ANSI/ISA-84.01-1996 to promote compliance with the PSM regulation. The International Electrotechnical Commission (IEC) created a similar document, IEC 61508, which is an umbrella standard that covers numerous industries. IEC standard 61511 is the process-sector specific standard that falls under the IEC 61508 umbrella. This standard, when ratified, will be reviewed by ISA SP84 and accepted as a replacement for ANSI/ISA-84.01, possibly with some modification. The IEC standard 61511 will have a global scope.

ANSI/ISA-84.01-1996 and IEC 61508/61511 use the concept of the *safety life cycle* as a tool for managing the application of safety instrumented systems. As an integral part of the safety life cycle, the selection of an SIL forms the foundation of a management system that can assure safe processes. International standards for SIS design, such as ANSI/ISA-

84.01-1996 and IEC 61508 and 61511, require that an SIL be selected. These standards are the basis of organizations' efforts to comply with the local and national laws and regulations that govern processes that contain significant risks. Many "authorities having jurisdiction," who are responsible for enforcing these laws and regulations, tend to view complying with such international standards as equivalent to complying with "good and generally recognized engineering practices" clauses.

1.1 Safety Integrity Level

Safety integrity levels (SILs) are categories based on the *probability of failure on demand* (PFD) for a particular *safety instrumented function* (SIF). The categories of PFD range from one to three, as defined by ANSI/ISA-84.01-1996, or one to four as defined by IEC 61508 and 61511. Table 1.1 shows the PFD ranges and associated risk reduction factor (RRF) ranges that correspond to each SIL.

Table 1.1 Safety Integrity Levels and Corresponding PFD and RRF

SIL	PFD Range	RRF Range
4	$10^{-4} \rightarrow 10^{-5}$	10,000 → 100,000
3	$10^{-3} \rightarrow 10^{-4}$	1,000 → 10,000
2	$10^{-2} \rightarrow 10^{-3}$	100 → 1,000
1	$10^{-1} \rightarrow 10^{-2}$	10 → 100

The SIL is the key design parameter specifying the amount of risk reduction that the safety equipment is required to achieve for a particular function in question. If an SIL is not selected, the equipment cannot be properly designed because only the action is specified, not the integrity. To properly design a piece of equipment, two types of specifications are required: a specification of what the equipment does and a specification of how well the equipment performs that function. The safety integrity level addresses this second specification by indicating the minimum probability that the equipment will successfully do what it is designed to do when it is called upon to do it.

In comparing safety equipment design to the more traditional design of a control system, one could say that specifying the action of a safety instrumented function and not specifying the SIL is like specifying a control valve without specifying the flow rate (or Cv) of the valve. Although you could pick a valve without knowing the flow rate (perhaps by simply choosing the same size as the piping and selecting equal percentage trim), your selection would not be optimal. You would have no guarantee that the valve would be able to pass the proper flow rate, and you would almost be guaranteed to have selected a valve that is oversized, and thus overpriced. You could improve performance and lower capital expenditures by investing the effort required to select a piece of equipment that not only performs the proper function, but also has the required performance characteristics.

Selecting safety integrity level involves giving a numerical target upon which subsequent steps in the safety life cycle are based. Thus SIL selection offers an important guide when you are selecting equipment and making maintenance decisions. The SIL is documented along with the SIS operational requirements and logic as part of the safety requirements specification. This specification provides the foundation for all of the safety life cycle activities an organization later conducts.

IMPORTANT: The process we are referring to as SIL selection in this book has been described by many other terms, including *SIL determination* and *SIL classification*. We specifically chose *SIL selection* because it describes the overall process most clearly. *Determination* is a vague term allowing too many variations in connotation. *SIL classification* implies that the process does not involve making a decision and that every situation is the same if you know its category. *Selection* is the clearest and most descriptive term because it emphasizes the act of choosing the correct value based on clear criteria.

1.2 Safety Instrumented Functions

In this book, we will adopt the terminology of IEC 61511, wherein a safety instrumented function (SIF) is an action a safety instrumented system takes to bring the process or the equipment under control to a safe state. This function is a single set of actions that protects against a single specific hazard. A safety instrumented system (SIS), on the other hand, is a collection of sensors, logic solvers, and actuators that executes one or more safety instrumented functions that are implemented for a common purpose, such as a group of functions protecting the same process or implemented on the same project. Note that the term *SIF* often refers to the equipment that carries out the single set of actions in response to the single hazard, as well as to the particular set of actions itself. Here are some examples:

- SIF 1: High reactor temperature closes the two reactor feed valves.
- SIF 2: High column pressure or high column temperature closes a valve in the steam to the reboiler.
- SIF 3: High column pressure closes the two reactor feed valves.

The logic for all safety functions is performed in a safety PLC. This PLC would then combine with all of the equipment associated with each SIF to constitute the SIS.

Figure 1.1 Safety Instrumented Functions versus Safety Instrumented Systems

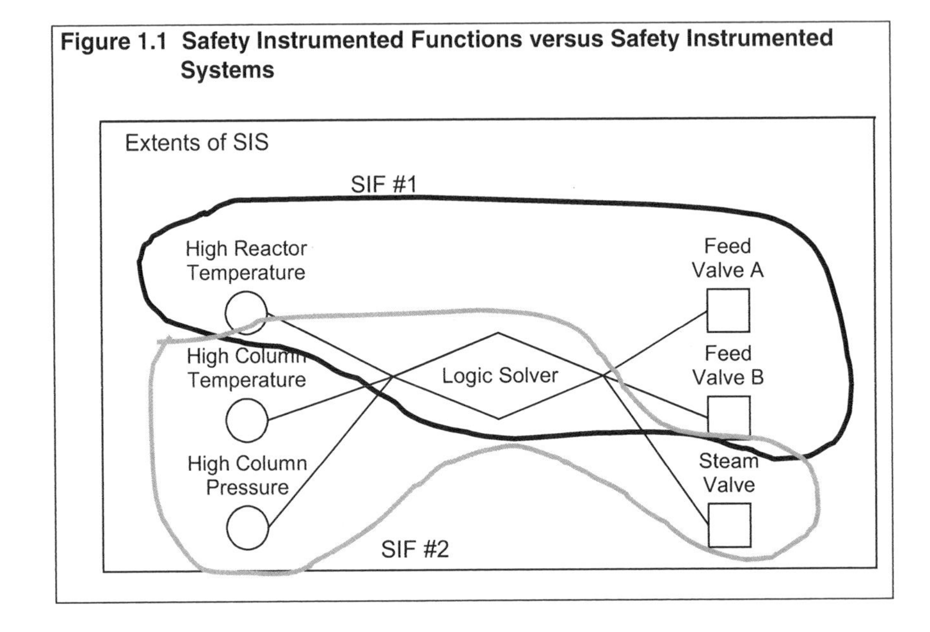

You may implement one or more SIFs in a SIS, as shown in figure 1.1. ANSI/ISA-84.01-1996 uses the terms *SIF* and *SIS* in a somewhat interchangeable and confusing way. IEC 61511 makes the distinction between SIF and SIS very clear. As figure 1.1 shows, a safety function can include multiple inputs and outputs. SIF 1 is executed with two outputs, that is, the two reactor feed valves, and SIF 2 has two inputs, that is, the high pressure and high temperature measurements. It is also important to note that a multiple SIF system can include common equipment. For instance, in figure 1.1, both SIFs use the same logic solver. In instances where common equipment is used in multiple SIFs, the common equipment item should be designed to meet the SIL of the SIF that has the highest requirements.

IMPORTANT: The SIL belongs to the specific safety instrumented function (SIF), not to the entire safety instrumented system (SIS). When an equipment item is common to multiple SIFs, it should be designed to meet the highest SIL requirements of the SIF it supports.

Throughout this book, we use the word *selection* to describe the overall process of choosing an SIL and *assignment* to define the final stage of the process, in which the SIL is assigned based on the results of the analysis that led to the selection.

1.3 SIL Selection and Risk

The reason an organization should use a systematic methodology, which includes layer of protection analysis, to select safety integrity level is to make the choice that best reduces risk. A good decision during this phase of the safety life cycle will ensure that the safety system specified will be cost-effective while still providing appropriate loss prevention. To make the best decision about safety integrity level, an SIS designer needs to completely understand not only the potential likelihood of an unwanted event, but also the possible consequences of that event. Viewing either of these two facets of the risk equation in isolation will yield poor results. Once the risk is known, one must determine how to reduce that risk to a tolerable level. The amount of risk that an organization is willing to tolerate will determine the amount of risk reduction it needs. Many proposed risk reduction projects are financially impractical because the amount of risk reduction they provide is grossly disproportionate to their cost. SIS designers must weigh the amount of risk reduction an SIF achieves against the equipment's cost. Good designs will optimize the return on investment.

Since safety integrity level is defined by the amount of risk reduction an SIS provides, it is important to understand what is meant by risk. There are many different types of risk, and it means different things to different people, but risk has a particular meaning in the context of SIL selection. Here we define risk as a measure of the likelihood and consequences of adverse effects when a process goes out of control and its hazards are realized. Risk is the product of both *likelihood* and *consequence*. The total risk can only be known when both the likelihood and consequences are known. Knowledge of either in isolation is simply not enough to properly solve risk reduction problems.

The risk reduction process is illustrated in figure 1.2. Before selecting an SIL, you must evaluate the inherent risk of the process. The starting point for SIL selection, or the baseline risk, is the level of risk that exists after considering all non-SIS mitigation measures (e.g., relief valves, dikes). Once the baseline risk is determined, you can employ an SIS to further reduce the risk by decreasing the likelihood of an incident. Each additional SIL step, by definition, reduces the likelihood of harm by an order of magnitude. For example, if the baseline likelihood were 10^{-2} per year, an SIL 3 system would reduce the likelihood by three orders of magnitude to 10^{-5} per year. The appropriate SIL is the one that reduces the risk to a tolerable level, or *as low as reasonably practicable* (ALARP). The ALARP concept is explained in more detail in section 3.1. In figure 1.2, SIL 1 would be appropriate if the cost of SIL 2 could not be reasonably justified. SIL 2 is acceptable without further analysis so the cost effectiveness of SIL 3 need not be investigated. This is because the gen-

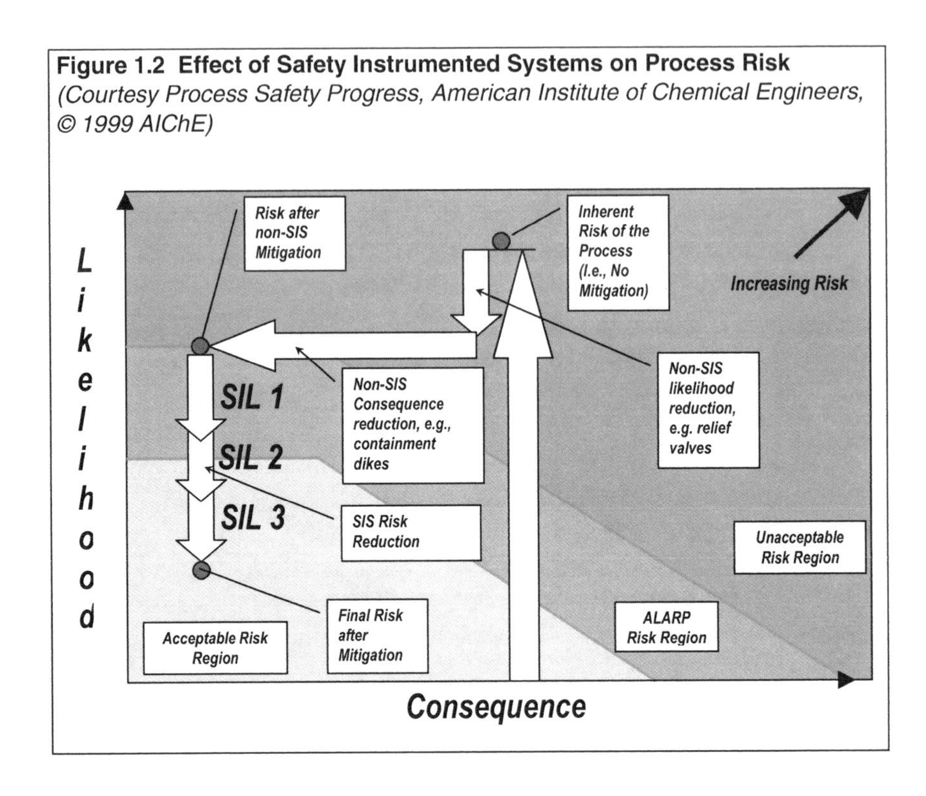

Figure 1.2 Effect of Safety Instrumented Systems on Process Risk
(Courtesy Process Safety Progress, American Institute of Chemical Engineers, © 1999 AIChE)

eral acceptability of SIL 2 directly implies that the additional SIL level could not be cost-justified.

1.3.1 Consequence

If a process goes out of control, the potential energy contained in the process, in the form of temperature, pressure, and chemical reactivity, may be released and cause harm. This harm is the result of the impact of the event on various receptors. The consequence of a process incident could result in impacts to one or more receptors, such as:

- People
- Property
- Environment
- Business (production and profits)

Estimating the magnitude of the consequence is the first step in the process of selecting an SIL. Many consequences are small enough to be acceptable regardless of how often they occur. If this is the case, the analysis can stop at this stage. Consequence analysis is typically performed by risk analysts in the loss prevention department of an organization or by consultants, especially if the consequence is due to the release of a flammable or toxic chemical. Understanding the consequences of these releases requires a thorough understanding of analytical chemistry, physics, and meteorology. Such knowledge is often beyond the scope of control system and process engineers. We consider consequence analysis in more depth in chapter 6.

The practice in industry has been to limit the scope of risk analysis to such consequences as the potential for injury to workers and the public as well as environmental impact. This scope is required by the mechanical integrity clauses of the U.S. OSHA and EPA process safety management (PSM) and risk management program (RMP). However, the potential financial losses resulting from property damage and business interruption are often just as significant, especially in the case of fires and explosions. If an organization addresses only risks to safety and environment, it loses from its analysis the financial benefits it could gain from reducing risks to property and business. Omitting these two impacts may cause a company to select a lower (less demanding) SIL, and thus fail to achieve significant, long-term benefits.

1.3.2 Likelihood

The likelihood of harm from a process hazard is the combination of the frequency and probability of all contributing events. It is unlikely that a single point failure will result in a hazardous consequence, since most process plants are designed with multiple layers of protection. For example, the analysis of the series of events leading to harm could involve the following:

- Frequency of process failure
- Probability of safeguards failing, resulting in a hazardous release
- Probability that the release will result in a harmful impact

To determine the overall likelihood of harm, you must divide the sequence into basic events that are narrow enough in scope that historical experience will yield a reasonably good estimate of their frequencies and probabilities. The level of detail for a likelihood analysis can range from a simple qualitative estimate (based on engineering judgment) to a quantitative analysis that uses sophisticated fault propagation modeling techniques in combination with historical data. These procedures are discussed further in chapters 7 and 8.

1.3.3 Tolerable Risk and SIL Assignment

The United States does not require companies to use formal risk decision criteria. However, company-based criteria can help ensure that an SIS achieves the desired level of risk reduction. Many companies have adopted internal risk guidelines to ensure an appropriate level of protection for workers and the surrounding community. Often, government and corporate risk guidelines only set limits for safety (i.e., potential fatalities and injuries) and environmental impacts, while ignoring the financial risks to property and production. We consider these government and corporate risk guidelines in more detail in chapter 3.

The final step in the SIL selection process is to compare the risk that is estimated to exist in the process with the amount of risk an organization is willing to tolerate. The difference between the existing risk and the tolerable risk determines the amount of risk that must be reduced. As mentioned earlier, SIL is a measure of required risk reduction. Once the required risk reduction is known, it can be converted into a probability of failure on demand (PFD). This is the PFD required of the safety instrumented function, and thus its required safety integrity level.

1.4 Qualitative versus Quantitative SIL Selection

The standards governing SIS design require that SILs be selected, but they do not place any requirements on the methods that should be used for that selection. As a result, many SISs are either overdesigned or underdesigned because poor risk analysis techniques used during the SIL selection process resulted in incorrect SIS design specifications. Qualitative techniques are often used to determine the risk upon which the selection of an SIL is based. A company's overall risk analysis is much more accurate when it uses layer of protection analysis (LOPA) to improve the accuracy of its likelihood estimate and when it uses quantitative consequence modeling. When the existing layers of protection are accounted for more accurately, the integrity requirements of the SIS can be relaxed. As a result, the SIS has both lower initial installation costs and lower overall life cycle costs. This accurate accounting also provides the organization with a clearer understanding of what contributes to the risk reduction and how. This knowledge can then lead to more informed decisions regarding what SISs and what non-SIS risk reduction methods are most effective and appropriate for the situation at hand.

There are two general ways to perform risk analysis: qualitatively and quantitatively. *Qualitative methods* use engineering judgment and personal experience as the basis for making decisions. Often, qualitative methods use rules and checklists to determine how processes should be designed. These rules and checklists are developed by a group of engineers over many years, and are based on a sort of "institutional memory" of the accidents and near misses that have occurred at a plant. Although qualitative methods are based on experience, there are no precise numerical records to verify the professional judgment of the qualitative analyst.

Quantitative methods strictly use historical data and mathematical relationships to estimate risk. These tools and techniques are based on scientific analysis and data obtained from recorded observations. Decisions arrived at using quantitative tools are based on the actual risk that a process poses, not on rules about preventing past accidents. Quantitative decisions are more objective than qualitative decisions, and make it easier to compare alternative designs because they provide objective, measurable criteria.

Both methods have their strengths and limitations. Qualitative methods are effective in situations where the process has a long history and its risk reduction techniques are well established. Quantitative methods are more objective and allow the proactive application of risk reduction to novel situations.

1.4.1 Problems with Qualitative Techniques

The risk analysis methods that organizations use today to select SILs rely on either qualitative or quantitative techniques or, in some cases, a combination of the two. Qualitative techniques are very subjective and can be susceptible to large errors because of the psychological traps that can cloud human thought.

Table 1.2 Common Psychological Traps of Qualitative Analysis

Relying too much on first thoughts ➔	The anchoring trap
Focusing on the current design ➔	The status quo trap
Protecting earlier choices ➔	The sunk-cost trap
Focusing on dramatic events ➔	The recallability trap
Neglecting relevant information ➔	The base rate trap
Slanting probabilities and estimates ➔	The prudence trap

Researchers have repeatedly shown that humans are actually quite horrible judges of the frequency of events that occur at long intervals. Despite this, human experts are expected to evaluate the difference between two events whose frequency is less than once in one thousand years (or ten lifetimes)! It's no wonder their results are often poor.

Luckily, the recallability trap and prudence trap conspire to create risk estimates that are too conservative rather than too aggressive. Although conservative estimates may mean the part of a plant in question is judged to be slightly safer than it really is, the resulting overdesigned systems require much more capital to install and maintain than is necessary. Studies have shown that more than 50 percent of a typical refinery's safety functions are overengineered. The extra capital spent for marginal improvement in a few arbitrary areas could always be spent more wisely elsewhere in the plant to improve safety more equitably over a broader range of situations. It is thus safer and more efficient to look at the entire plant consistently and objectively to ensure that all systems meet the same requirements based on the best information available. In one specific case where an organization took these benefits in financial terms, lowering unbalanced integrity requirements for a single refining plant to more realistic, more consistent levels lowered operating expenditures by more than $100,000 per year.

1.4.2 Improving Likelihood Estimation with Quantitative LOPA

To obtain sufficiently accurate risk estimates, one should ensure that the risk analysis process is more quantitative. One can do this by evaluating the existing layers of protection using a fault propagation modeling technique called *layer of protection analysis* (LOPA). Using LOPA will enable the analyst to estimate quantitatively the frequency of unwanted events instead of relying on purely qualitative techniques. LOPA provides the following benefits:

- It lowers the life-cycle cost of SIS.
- It supplements existing procedures (e.g., hazard matrix, risk graph). Adding LOPA does not require an organization to completely change other procedures.
- It is neither excessively complicated nor time consuming.
- It is more accurate than purely qualitative estimates.
- Its tools and data are commercially available and inexpensive.
- It identifies potential gaps in the plant's overall protection that can often be fixed very economically.

Using LOPA to estimate likelihood complements almost every SIL selection protocol currently in use, from Hazard Matrix all the way to Quantitative Risk Analysis. LOPA is accomplished by performing the following steps:

1. Define the unwanted impact.
2. Determine and list all of the initiating events that can cause the unwanted impact.
3. Determine and list all the layers of protection that are available for preventing an initiating event from propagating into the unwanted impact (the protection layers for each initiating event may be different).
4. Quantify the frequency of the initiating event(s), based on historical data and engineering judgment.
5. Quantify the effectiveness of the layers of protection in terms of probability of failure on demand (PFD), based on historical data and engineering judgment.
6. Calculate the resulting frequency of the unwanted impact.

1.4.3 Improving Consequence Estimation with Quantitative Modeling

Quantitative analysis of the consequences of chemical releases is currently even less prevalent than quantitative modeling of likelihood. The reason for this is a lack of familiarity with consequence models and the physical phenomena surrounding chemical hazards. Consequence modeling for chemical releases is an exercise in applied chemistry, heat and mass transport, and meteorology. The skill set required to perform this type of analysis is typically not very strong in a traditional control systems engineer, whose background is usually either electrical or mechanical engineering. As a result, most attempts to quantify risk are strictly exercises in probability and statistics, which are unfortunately insufficient for properly analyzing chemical releases.

Another reason chemical release modeling is not frequently used is that it is difficult to perform, even for professional risk analysts. The number of parameters that can affect the consequence of a chemical release is staggering. This problem is compounded by the fact that small changes in seemingly minor parameters can have very large effects on the results of an analysis. The sheer number of scenarios that must be analyzed also complicates the process. There are quite a few software packages available that perform quantitative consequence analysis on chemical releases. However, most are difficult to use and expensive, and require a great deal of risk analysis expertise not only to use but to interpret their results.

Although quantitative consequence analysis is often difficult today, it is becoming easier on several fronts. The EPA's Risk Management Plan (RMP) rule in the United States requires owners/operators of chemical plants to calculate the consequences of a worst-case release of the hazardous chemicals on their site. They must then use this calculation to determine if offsite consequences are possible and to estimate the number of persons and sensitive environmental receptors that could be impacted. In making this requirement, the EPA itself was required to develop a simplified protocol that would allow a small process owner, with no risk analysis staff or experience, to perform the Offsite Consequence Analysis (OCA). The EPA developed an *Offsite Consequence Analysis Guidance* that simplified the process of quantitatively modeling chemical releases by standardizing a large number of parameters to define a "worst case." The EPA also tailored equations and look-up tables so that it would be easy to determine consequence zones, requiring only a few easy-to-determine parameters such as release quantity and temperature. Private industry has also expanded upon the EPA's framework, and new software tools based on the EPA's OCA procedures are becoming available.

Quantitative chemical release modeling, like layer of protection analysis, complements most existing SIL selection protocols well. Organizations can use the results of the analysis either directly in risk calculations or to support the selection of a qualitative category when using qualitative methods such as hazard matrices and risk graphs.

1.5 Benefits of Systematic SIL Selection

Selecting safety integrity levels systematically using quantitative methods yields a number of benefits:

1. ***Lower risk*** – The methods described in this book help the user determine an appropriate level of risk reduction and help ensure that systems are not overdesigned. By ensuring that SISs are not overdesigned, these methods enable scarce risk reduction resources to be distributed more evenly. This means more risk reduction projects can be completed with the same resources, and lowering the overall level of risk in the plant.

2. ***Understanding of the hazards of the process*** – Using this systematic process to select SILs enables process owners to more rigorously analyze the hazards in their plants. This knowledge is critical during every phase of the safety life cycle. When qualitative methods are used, SISs can be designed faster, but important properties of the hazards may not be considered. This can cause poor design and operation decisions later in the safety life cycle.

3. ***Maximum return on risk reduction investment*** – As mentioned earlier, the key to an effective risk management program is not maximizing the amount of risk reduction each protection layer provides. Rather, it is selecting an appropriate amount of risk reduction, so that risk reduction resources are optimally distributed. Making minor risk reductions that are costly to implement is not in the best interest of the organization.

4. ***Making the right choices for the situation at hand*** – Determining how much risk reduction is required is essentially a decision process that yields a multitude of answers, many of which are correct. Process owners must make the selection that is right for their specific process. By using the tools and techniques presented in this text you can be comfortable that the decisions you make will provide excellent results. Our process is designed to defeat second-guessing, strengthen management buy-in, and create a trail of documentation that will justify your decisions.

5. ***Compliance with standards*** – Both national and international standards require organizations to select safety integrity levels and quantitatively verify that they have been achieved. These standards also form the basis of "good and generally accepted engineering practices," which many regulatory agencies use to determine compliance with process safety laws and regulations.

1.6 Objectives of this Book

This book is intended to prepare the reader to select SILs systematically by using a variety of tools and techniques, both qualitative and quantitative, and by applying the most appropriate protocols to each problem. To do this, we break the subject of SIL selection into individual skills. We develop each of these areas separately, and then bring the entire selection process together through a comprehensive case study. The individual subjects and skills that the reader will develop are the following:

- *Understand the safety life cycle concept and the context of SIS application.*
  The concept of a safety life cycle has been described in several important national and international standards, such as the ANSI/ ISA-84.01-1996, the IEC 61508, and IEC 61511. This concept develops a global context for specifying, designing, implementing, and maintaining safety instrumented systems so as to achieve overall functional safety. SIL selection and the analysis techniques that support this process are a key part of this safety life cycle process.
- *Understand risk and the application of tolerable risk guidelines.*
  Risk is a component of everyday life. Virtually no tasks are completely free of risk. The important point is to balance risk and reward in an equitable way that meets moral, legal, and financial obligations.
- *Understand and be able to apply the rules of probability.*
  The rules of probability form the foundation of most risk analysis tasks. Before any understanding of the hazards that processes pose can be developed, probability rules must be known. In addition to basic probability theory, the book will present fault tree analysis as a way to demonstrate and determine the failure probabilities of complex systems.
- *Understand the range of consequences of process accidents and how to estimate them.*
  Process accidents can produce a range of consequences, from no significant impact up to many fatalities. It is important to clearly understand the consequence of an accident when making decisions about its tolerability.
- *Understand how to use event trees to estimate the probability of uncertain events.*
  Sometimes the probability of an unwanted event will be very difficult to determine either qualitatively or through traditional statistical methods. In these situations, one must model the accident

probability based on the rates of initiating events and the effectiveness of the protection layer. Event tree analysis is a proven and effective method for modeling the probability of complex situations.

- *Understand the mechanics of layer of protection analysis.*
  The first step in understanding and being able to employ LOPA is to know its underlying mathematics and its relationship to other fault propagation modeling techniques. LOPA is a variation of event tree analysis. Once the relationship between the two modeling forms is described, the basis for doing LOPA calculations becomes apparent.

- *Be able to identify and quantify events that can initiate an accident.*
  After understanding the mechanics of LOPA, the science and art of determining the structure of the LOPA analysis comes next. All LOPA and event tree studies begin with initiating events that, if not stopped by a layer of protection, will propagate into an accident. Identifying the initiating events that can cause accidents is crucial for identifying areas where the process would benefit from a safety instrumented system and for building the structure of the LOPA analysis. In addition to building the LOPA structure, you must quantify the initiating events to determine the overall accident frequency.

- *Be able to identify and quantify protection layers that prevent accidents.*
  Understanding the layers of protection is critical to being able to estimate event frequency. The effectiveness of each layer of protection is a function of its mean time to failure (MTTF), test interval, and other variables. This book explains the variables that impact the effectiveness of protection layers and describes how to manipulate measurements and equations to demonstrate effectiveness in the desired format.

- *Be able to assign SIL targets using various representations of tolerable risk.*
  Applying the chosen tolerable risk levels is an important, but not always straightforward, task. If tolerable risk levels have been stated using qualitative methods, applying them is even more subjective. This final section explores the various methods used in industry to represent tolerable risk, and demonstrates how they are used for SIL assignment.

1.7 Summary

This chapter addressed the general purpose and limitations of selecting safety integrity levels. The SIL describes the amount of risk reduction that a safety instrumented function should achieve. The SIL an organization selects should reduce risk from its initial unacceptable level to a defined tolerable level and maximize the return on investment for the risk reduction project. SIL is defined as the amount of risk reduction required to make the risk of a process tolerable. SILs are the categories of the probability of failure on demand (PFD) that an SIS must achieve. These categories are defined in international standards that describe SIS design. Selecting safety integrity levels is a necessary step in a complete definition of a safety instrumented function. This selection is also required by international standards as part of the safety life cycle.

Risk is the product of the likelihood and consequences of an unwanted event. To understand risk, one must analyze both components of the risk equation. Risk reduction decisions can be made either qualitatively or quantitatively. Qualitative methods are rule-based and reactive. They use a large knowledge base of prior accidents and near misses to determine how risk reduction should be employed. Quantitative methods, on the other hand, use scientific methods and records of historical operation to calculate risk. Quantitative methods are proactive, objective, and risk-based.

Many SIS designs suffer from excessive integrity requirements caused by poor risk analysis techniques. Analysis techniques that provide a more realistic representation of process risk will yield systems with integrity requirements that are not overengineered. As a result, organizations can realize huge savings in the life cycle cost of safety systems without sacrificing overall safety.

Layer of protection analysis (LOPA) is a fault propagation modeling technique that allows the user to estimate the frequency of rare events based on the likelihood of initiating events that contribute to the final outcome. LOPA is used to improve the accuracy of estimating likelihood; as a direct result, the overall risk analysis is more accurate. LOPA can be incorporated into existing procedures without excessive effort, complicated procedures, or high-priced consultants. The systematic selection of SILs using layer of protection analysis offers the benefits of lower overall plant risk, a more thorough understanding of the hazards posed by the process, maximization of the return on the organization's investment in risk reduction, and assurance that the best choices have been made.

1.8 Exercises

1.1 What range of probability of failure on demand (PFD) is associated with SIL 2? What range of risk reduction factors is associated with SIL 1? How are risk reduction factors and probability of failure on demand related?

1.2 Name one of the standards that describes the safety life cycle for safety instrumented systems.

1.3 Explain the difference between a safety instrumented function and a safety instrumented system.

1.4 What are the two components of risk?

1.5 Describe three of the psychological traps that make qualitative estimates of risk inaccurate.

1.6 Which part of the risk equation (i.e., consequence or likelihood) is layer of protection analysis used to estimate?

1.9 References

1. ANSI/ISA-84.01-1996 - Application of Safety Instrumented Systems for the Process Industries. Research Triangle Park, NC: ISA, 1996.
2. Center for Chemical Process Safety of the American Institute of Chemical Engineers. *Guidelines for the Safety Automation of Chemical Processes.* New York: Center for Chemical Process Safety of the American Institute of Chemical Engineers, 1993.
3. Hammond, J. S., H. Keeney, and H. Raiffa. *Smart Choices – A Practical Guide to Making Better Decisions.* Boston: Harvard Business School Press, 1999.
4. International Electrotechnical Commission. "Functional Safety of Electric/Electronic/Programmable Electronic Safety-Related Systems," IEC 61508. Geneva: IEC, International Electrotechnical Commission.
5. Marszal, Edward M. "Layer of Protection Analysis – Decreasing Safety System Costs by Analyzing Layers of Protection." White paper, http://www.exida.com, October 2000.
6. Marszal, Edward M., Brad A. Fuller, and Jatin N. Shah. "Comparison of Safety Integrity Level Selection Methods and Utilization of Risk Based Approaches." *Process Safety Progress* 18, no. 4 (winter 1999).
7. Wiegerinck, J. "Design Philosophy E/E/PES – Classification and Implementation of Instrumented Protective Functions – The Shell Approach." Notes of presentation delivered at INTERKAMA, October 1999.

CHAPTER 2

Safety Life Cycle Context for SIL Selection

The purpose of a safety instrumented system (SIS) is to reduce the risk of a hazardous process to a tolerable level. Although selecting a safety integrity level (SIL) is vital to this purpose, an organization that wants to achieve what the standards call "functional safety" must expend significant additional effort over a broad range of other supporting safety activities. The ultimate goal of any organization's safety effort is then to execute these activities so as to achieve the desired level of safety as efficiently and effectively as possible. Governmental safety regulations and international standards all support this goal, with varying degrees of clarity. One area of strong clarity in all of the various standards and regulations is the definition of an overall safety life cycle or SLC. This chapter provides an introduction to the safety life cycle concept and identifies how the key aspects of SIL selection fit into this overall process.

2.1 Standards and the Safety Life Cycle

The concept of a safety life cycle has been put forward in several key national and international standards, such as ANSI/ISA-84.01-1996, IEC 61508, and IEC draft 61511. The SLC is essentially a method or process that provides a global context for specifying, designing, implementing, and maintaining safety instrumented systems so as to achieve overall functional safety in a documented and verified way. IEC 61511 defines the SLC as the "necessary activities involved in the implementation of safety instrumented function(s), occurring during a period of time that starts at the concept phase of a project and finishes when all of the safety instrumented functions are no longer available for use." This definition covers all of the aspects of an SIS from conception to cremation. The SIL selection and analytical techniques that support this process are a key part of the SLC in each of the relevant standards.

The need for a more formally defined SLC process has emerged over the past twenty years, as more complex electronic and programmable safety systems have come into common use. One early effort at defining the SLC in the public sector was *Programmable Electronic Systems in Safety-Related Applications*, parts 1 and 2, published by the U.K. Health and Safety Executive in 1987. This document focuses on programmable systems, but presents several more general qualitative and quantitative methods for evaluating systems well as design checklists.

In the 1990s the International Electrotechnical Commission (IEC) developed the IEC draft 61508 standard and its later version, the current IEC 61508, "Functional Safety of Electric/ Electronic/ Programmable Electronic Safety-Related Systems." IEC 61508 was developed as an umbrella standard to cover numerous industries. The IEC has since developed industry-specific standards to support 61508, such as IEC 61511 for the process sector and IEC 62061 for the machinery and manufacturing industry.

In the United States, ISA—Instrumentation, Systems, and Automation Society developed ANSI/ISA-84.01-1996, "Application of Safety Instrumented Systems for the Process Industries," to promote compliance with the 1996 OSHA Process Safety Management regulation. With the recent trend toward globally recognized standards, ANSI/ISA-84.01-1996 is expected to be replaced by the new IEC 61511 process-industry standard when the latter is ratified in its final form. Both the IEC and ISA standards present versions of a specifically designated safety life cycle designed to guide the implementation of safety-related or safety instrumented systems.

These standards are gaining wide acceptance and now form the basis for compliance with the local and national laws and regulations that govern processes that pose significant risks. Many of the authorities responsible for enforcing these laws and regulations view complying with international safety standards as equivalent to complying with "good and generally recognized engineering practices" clauses. Thus, understanding the overall SLC process should be a prerequisite for selecting a SIL for any safety-related system.

The ISA standard version of the safety life cycle is summarized in figure 2.1. Its primary characteristics can be broken down into three main phases. The first, the analysis phase, focuses on developing a proper understanding of what hazards are present in a given situation, what safety precautions and layers of protection are appropriate, what specific safety instrumented systems are required, and what SIL should be selected for those systems. The second, the realization phase, focuses on the design and fabrication of the SIS. The third phase, the operation phase, covers the start-up, operation, maintenance, and eventual decommissioning of the SIS. These phases encompass the entire life process of the safety system, from concept through decommissioning.

The SLC process presented in IEC 61508 is shown schematically in figure 2.2. As with the ISA SLC, the overall process can be divided into analysis, realization, and operation phases. Parts of the 61508 standard

Figure 2.1 ANSI/ISA-84.01-1996 Version of the Safety Life Cycle

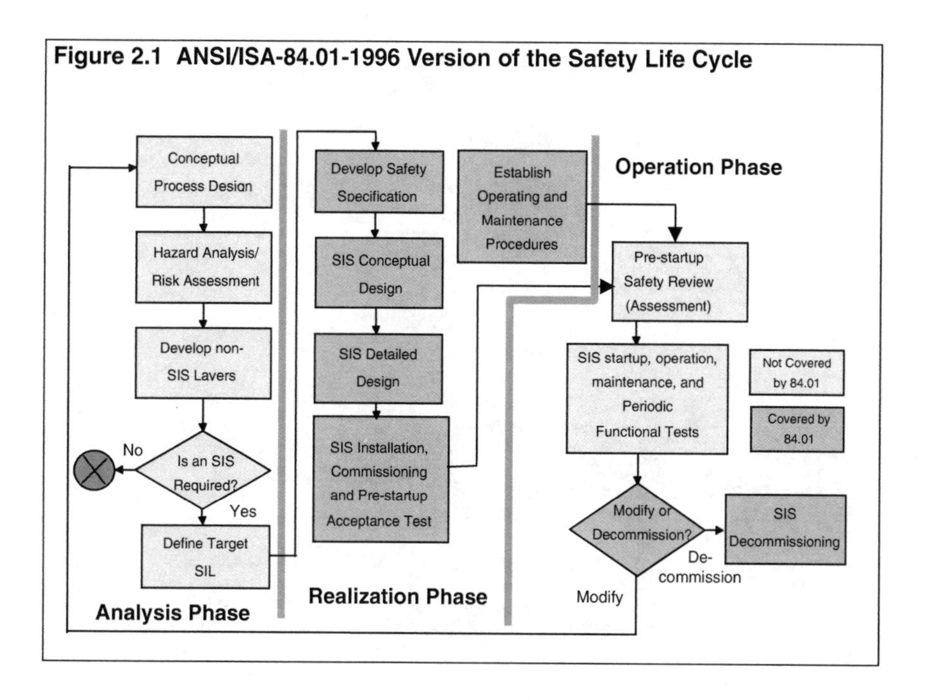

itself are generally structured along the same lines as the SLC, with different clauses in part 1 associated with each of the different steps numbered in the figure. IEC 61508 provides recommendations regarding the information required to execute each step as well as the output and documentation that should be produced in each step. However, the standard only includes general guidelines and recommendations on the life-cycle phases; it is not meant as a "cookbook" for functional safety.

Using IEC 61508 as the umbrella standard, IEC has developed IEC draft 61511 to provide specific guidance to the process industry. Although draft 61511 differs from 61508 and ANSI/ISA-84.01-1996, the overall analysis, realization, and operation phases are still clearly present (see figure 2.3). IEC draft 61511 also emphasizes the continuous functions of planning, management, assessment, and verification that support the sequential components of the life-cycle structure (see figure 2.3).

The essential aspects of analyzing, designing, verifying, and documenting are present in all of the different safety standards. However, within these broad categories, it is important that organizations understand where extra care is required to ensure that desired safety levels are achieved. For example, the U.K. Health and Safety Executive conducted a study of thirty-four accidents in different industries that were all caused by control and safety system failures. The results, shown in figure 2.4, indicate that the main cause of these accidents was errors in the system specification, including improper selection of SILs.

Figure 2.2 IEC 61508 Safety Life Cycle

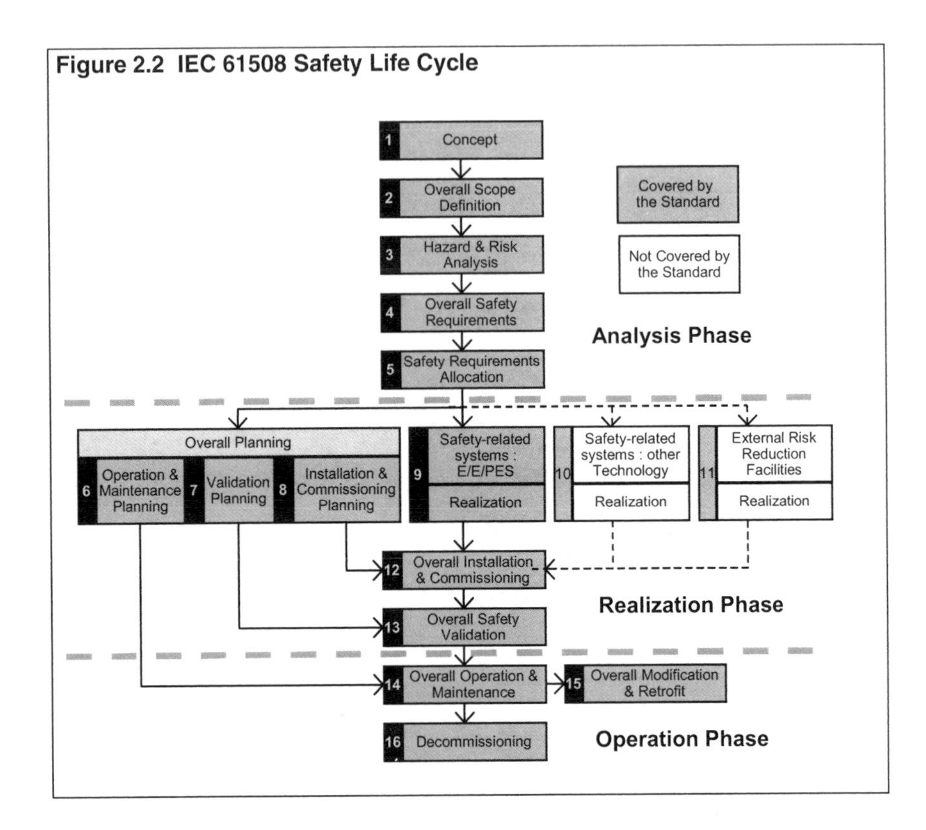

Figure 2.3 IEC 61511 Safety Life Cycle

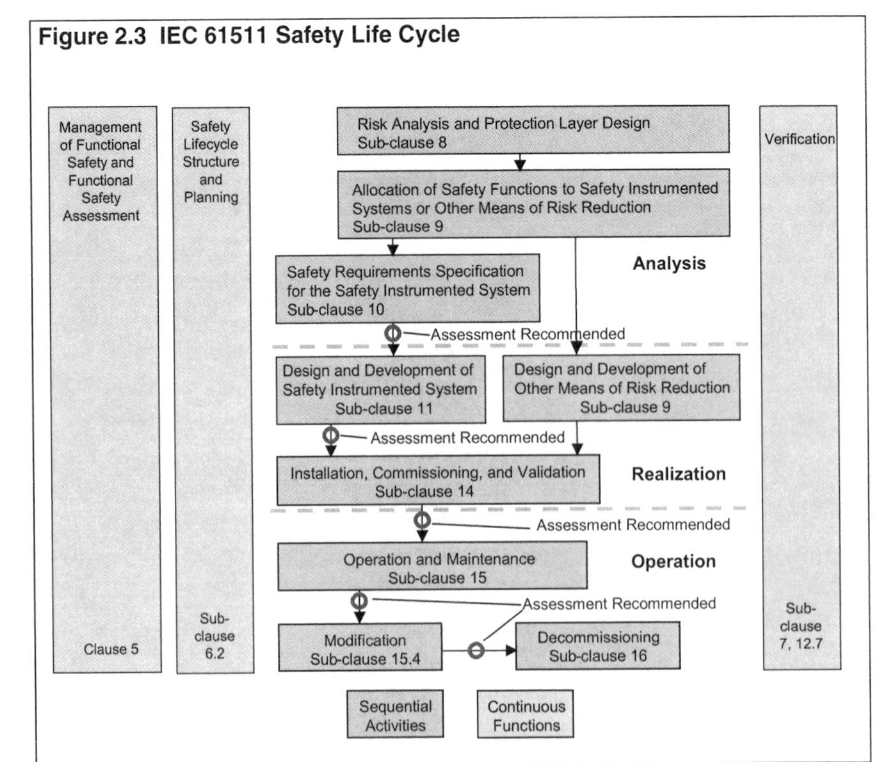

Figure 2.4 Effect of Safety Instrumented Systems on Process Risk

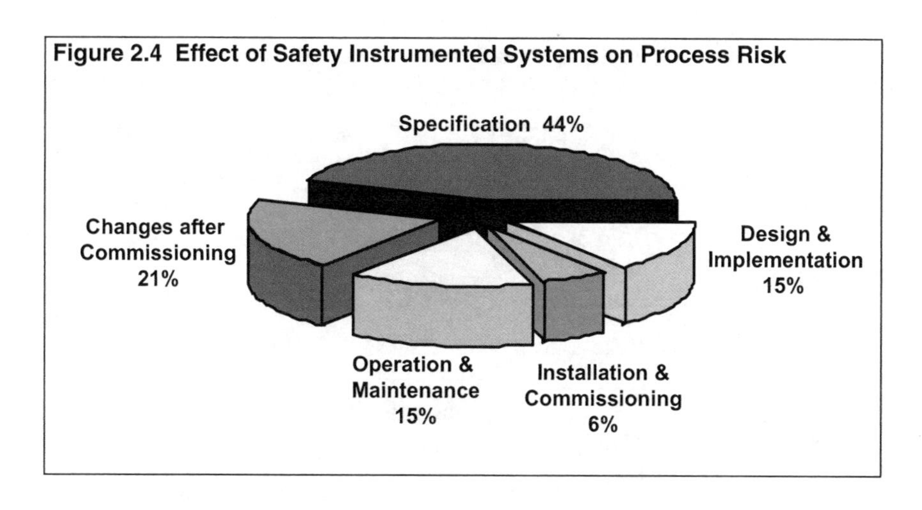

2.2 SLC Analysis Phase

The analysis phase of the SLC includes the initial planning, identification, and specification functions that are needed to properly apply safety systems to a process. The individual functions and the flow of information required to perform these tasks most effectively are summarized in figure 2.5. The analysis phase of the SLC focuses clearly on the SIL selection process and corresponds closely with the structure of the chapters in this book.

The SLC begins with the initial design of the process conceptually and the definition of the project's scope. It is important to clearly identify the project's purpose in terms of goals and measurable outcomes. If the project's initial definitions are ambiguous, team members can develop different versions of the project's scope and thus emphasize potentially conflicting aspects of the work. Clear definitions are particularly critical in projects such as capacity or production upgrades and new facility construction that have both operational and safety-focused objectives. Ideally, the organization should designate the relative importance, adequate resourcing, and proper scheduling of these objectives at the project's outset. Similarly, a single individual should be given ultimate responsibility for achieving both the safety-related and non-safety-related goals.

With respect to the components of the safety project, the organization's personnel should clearly understand the process and equipment under control. This understanding should include a preliminary idea of the potential process hazards and of the equipment and materials present, which will be developed further in subsequent steps. The organization should also understand the applicable regulations, laws, and standards, especially on the environmental side since in many cases, particularly in the United States, gaining environmental permissions is the long-lead, critical-path item. The scope definition should clearly desig-

Figure 2.5 Generalized Safety Life Cycle Analysis Phase

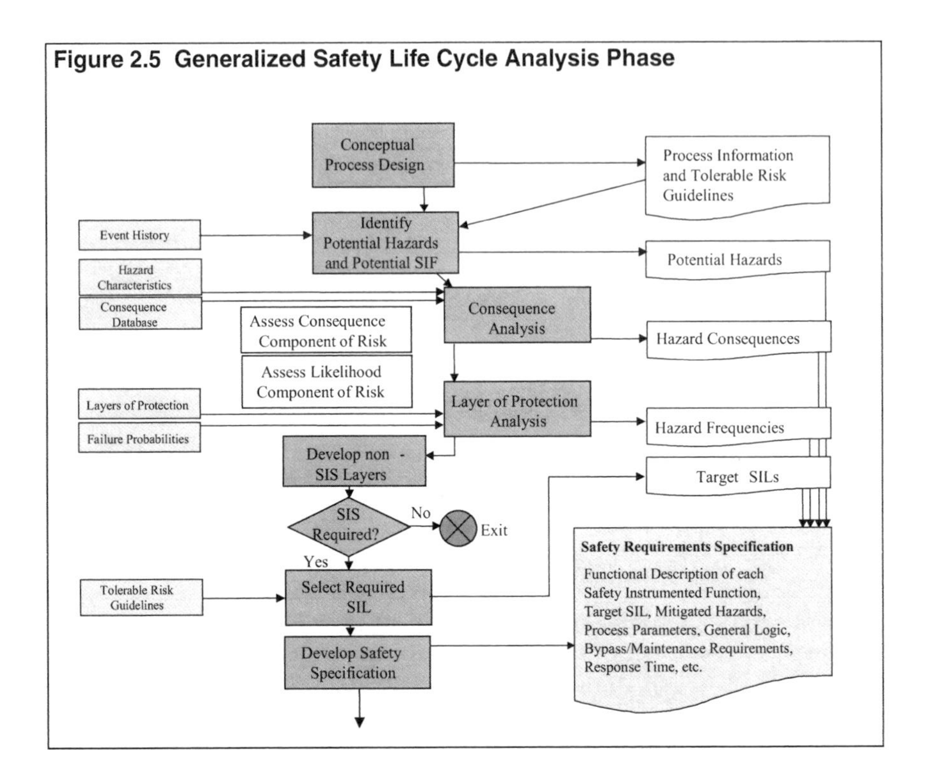

nate the limits of the process and equipment as well as the areas that will be addressed in later hazards and risk analysis steps.

Another issue that the organization should consider at the start of the project is the level of risk it will tolerate in its daily operation. This risk level should then be compared to the risks that are present in a process so the organization will know how much process risk must be reduced and what kind of safety equipment is needed to achieve this. (See chapter 3 for a more detailed discussion of tolerable risk.)

The next function in the SLC falls under the classification of hazard and risk analysis. As discussed in chapter 4, this analysis includes identifying the safety instrumented functions (SIFs) that may be needed to detect imminent harm and act to bring the process to a safe state. The first relevant task here is identifying the hazards (potential causes of harm) and the hazardous events that may potentially occur in the operation of the equipment or process. Many regulations, laws, and standards rigorously require this identification process by way of a process hazards analysis (PHA). (See section 4.1 of chapter 4 for a more detailed discussion of PHA.)

Once the organization has identified the hazards and potential SIFs, it needs to characterize the hazards in terms of both the magnitude of their consequences and the likelihood or frequency of their occurrence. The consequence analysis that is required to estimate the magnitude of the potential harm can be complex, depending on the hazard. This consequence analysis step is discussed in more detail in chapter 6. Analyzing the likelihood component of the risk (how often the harm might occur) involves understanding the different sequences of events that can lead to the harmful result. Chapters 7 and 8 provide information about the primary methods for calculating the frequency of potential accidents.

Related to these likelihood analysis methods is layer of protection analysis (LOPA), which is discussed in chapter 9. LOPA identifies and quantifies the non-SIS safety features of a process or equipment item, as well as any other factors that can prevent a harmful incident from occurring. By determining the probability that each of these "layers" can prevent the potential harm from occurring you can calculate the likelihood of the harmful outcome before considering the action of any SIF.

With information on the magnitude of the consequence, the likelihood of its occurrence, and the level of risk your organization can tolerate, one can determine whether an SIS is required to perform the SIF under consideration. If there is any difference between the risk present in the process after all non-SIS layers of protection have been credited and the risk that the organization can tolerate, it must be made up by using an SIS. The size of the risk gap or required risk reduction will determine which SIL should be selected for the SIF in question (see chapter 10).

Once one has identified and characterized all of the potential hazards, the risks they pose, any required SIFs, and their corresponding SILs, one must complete the analysis phase of the safety life cycle by documenting these efforts and results in the safety requirements specification (SRS). The purpose of the SRS, according to IEC 61508, "is to develop the specification for the overall safety requirements, in terms of the safety functions requirements and safety integrity requirements, for the E/E/PE safety-related systems, other technology safety-related systems and external risk reduction facilities, in order to achieve the required functional safety." The SRS documents exactly what is required if the safety system is to be designed, installed, and operated according to the subsequent life-cycle phases.

2.3 SLC Realization Phase

The realization phase of the safety life cycle encompasses the design, fabrication, installation, and testing of the SIS that was specified in the analysis phase of the project. The realization phase cannot be properly executed if the specification is not clearly and correctly developed from the results of the analyses conducted in the first phase of the life cycle. Figure 2.6 summarizes the individual functions and flow of information that are needed to perform the realization phase tasks most effectively.

Figure 2.6 Generalized Safety Life Cycle Realization Phase

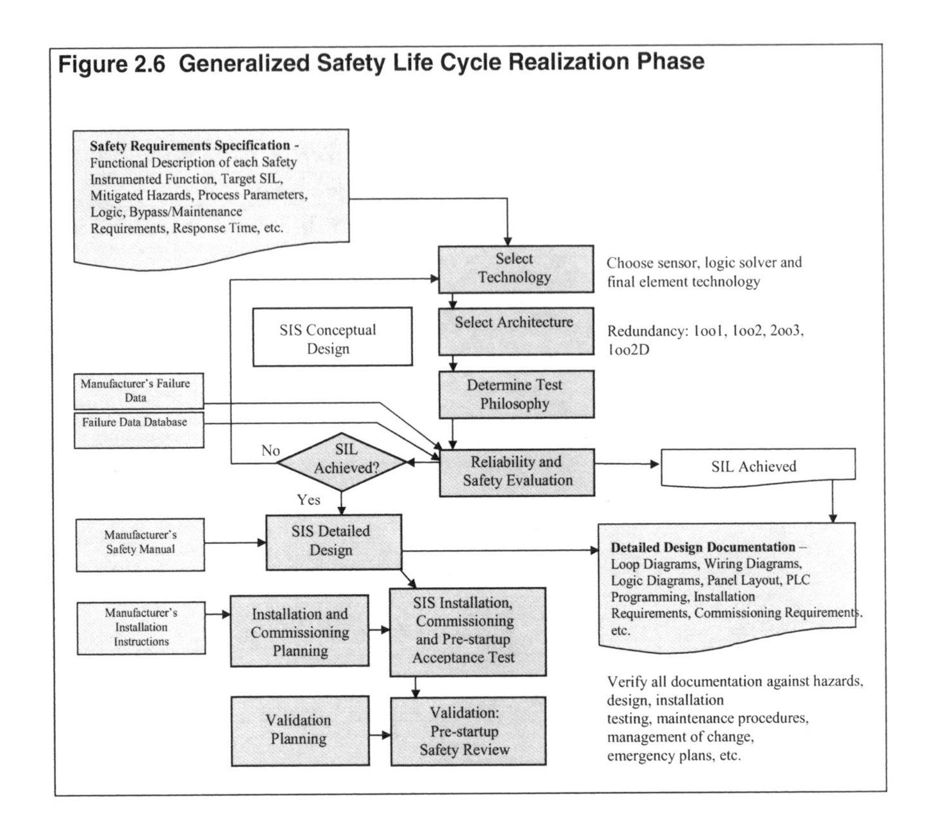

With the SRS in hand, the first task of the realization phase is to select what safety instrumented system technology and architecture are needed to meet the specification's requirements. The organization should address this conceptual design at the same time that it plans how the prospective system will be tested and verified to ensure it meets specification before being put into active use. A key part of this planning step is developing a maintenance and proof test schedule to ensure that one can find and repair any potential undetected failure in the safety equipment before the system is required to act. Both this proof test and the repair interval must be addressed properly since they affect the SIL for the system (see section 5.5.4 of chapter 5 for additional details).

Once the conceptual design is complete, the organization should analyze the prospective system to confirm that it meets the SIL that was selected and documented in the safety requirements specification. This analysis should include both the individual components of the system as well as the architecture used to configure the components. Only if the system can be shown to meet the selected SIL is it finally designed and fabricated. The detailed design of the SIS should be executed according to clear, established procedures. IEC 61508 presents additional hardware and software safety life cycles specific to the detailed design functions. (These other life cycles are outside the scope of this text, but further detail can be found in the 61508 standard itself.) As one would expect, the standards require appropriate documentation of both the SIL verification analysis and the detailed design of the SIS.

The final part of the realization phase is planning and executing the system's installation, commissioning, and validation. Once these tasks are finished, the SIS should be fully functional at the SIL that was selected to achieve a tolerable level of risk. With this, the SLC realization phase is completed. *Verification* and *validation* are two similar terms that describe the function of proving that a system will work. The terms are easily confused; even the IEC 61508 definitions of them are ambiguous. *Verification* has more of the connotation of an analysis or test, while *validation* suggests more the operation and functioning of the equipment in field situations. In addition to the original standards on the SLC realization phase, good references on the subject have also been published by the ISA and are listed in section 2.7 of this chapter.

2.4 SLC Operation Phase

The operation phase is the longest of the SLC phases. It begins at startup and continues until the SIS is decommissioned or redeployed, as shown in figure 2.7. The most significant part of this phase is the maintenance and testing of the SIS. As we noted in our discussion of the test philosophy and planning step, the system's SIL can be affected by the number of times it is tested and repaired to full functioning condition. A proper testing and maintenance regime begins with good planning and relies on solid documentation to show that the plan is being followed. Effective management of change is also important so any potential modifications to the system can be addressed properly. Depending on the exact nature of the modification, such change management should include a full return to the concept phase if circumstances warrant.

The decommissioning of the SIS ends the safety life cycle. Before the equipment is switched off for good, however, the organization should analyze the effects of the decommissioning on both the equipment or process directly under control and on any closely integrated systems.

2.5 Summary

In this chapter we presented the safety life cycle concept put forward in IEC and ISA standards. The SLC provides an overall context and general structure for applying safety instrumented systems, including the process of selecting SIL, the subject of this book. Although there are some differences in the details among the various standards, each of them clearly presents the SLC's analysis, realization, and operation phase functions.

Figure 2.7 Generalized Safety Life Cycle Operation Phase

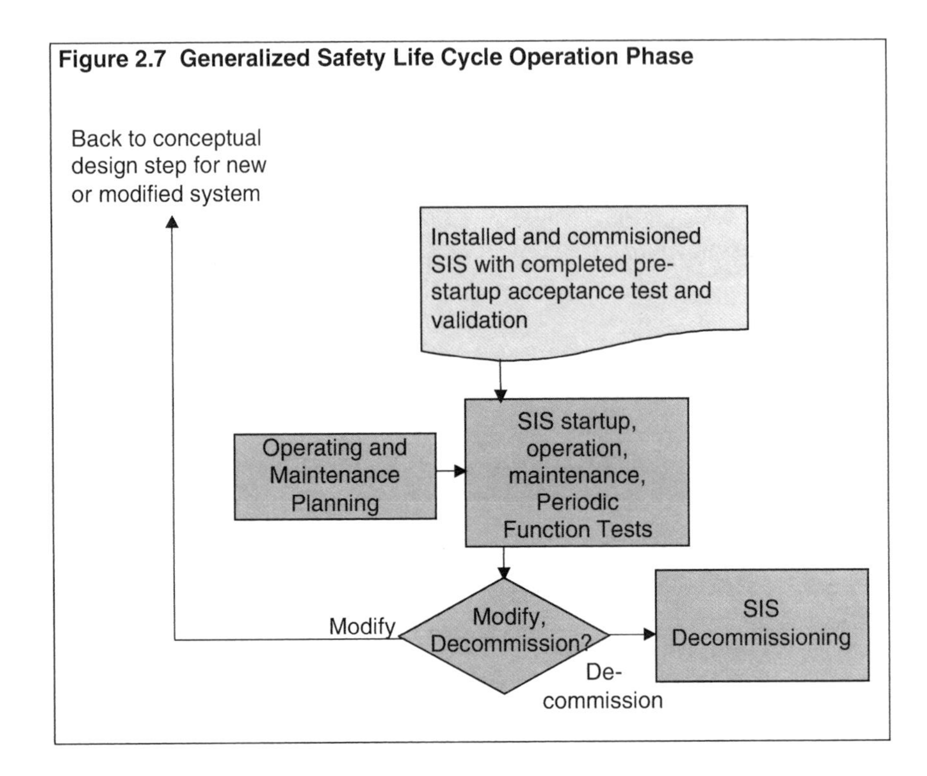

The SLC analysis phase focuses on the SIL selection process and runs from the initial conceptual design of the process and safety system through to the release of the safety requirements specification. With their emphasis on SIL selection, the steps in the analysis phase read like the table of contents of this book. Collecting supporting information on the process and on the level of risk that can be tolerated in the organization's situation leads directly into the analysis of process hazards. From the PHA, one identifies potential safety instrumented functions and then uses a combination of consequence and likelihood analysis, including LOPA, to characterize the amount of risk present without any SIS. If the risk is already within the level of tolerance, one need not employ any SIS and can rely completely on the other existing layers of protection. If the risk is intolerable, then you select the appropriate SIL to reduce that risk to an acceptable level and document the relevant information as part of the safety requirements specification.

The SLC realization phase focuses on designing and fabricating the SIS to meet the specifications yielded by the analysis phase. The realization phase initially involves selecting both the correct technology and architecture as well as the correct testing philosophy to achieve the required SIL. Then, after verifying that the proposed system will indeed meet the specification, one begins the detailed design and programming work. To complete the realization phase, one must plan, perform, and document the installation, commissioning, and validation functions before the SIS and the process are ready to operate.

The SLC operation phase runs from startup to decommissioning—the entire working life of the safety system. This phase is dominated by the activities of maintaining and function testing the system to ensure that it remains functionally safe. The operation phase ends either when the system is fully decommissioned and taken out of service or when it is modified according to a clear management-of-change procedure, which can start the life cycle all over again.

2.6 Exercises

2.1 What are the main steps in the analysis phase of the IEC 61508 safety life cycle (SLC)?

2.2 According to the U.K. Health and Safety Executive, the most harmful accidents are caused by errors to what part of the safety life cycle?

2.3 What should be included in a good safety requirements specification (SRS)?

2.4 Why should you consider the testing philosophy of the safety equipment before the detailed design of the safety system?

2.5 When is the function of the physical SIS tested during the safety life cycle?

2.6 What constitutes the key part of the operation phase of the SLC?

2.7 References

1. ANSI/ISA-84.01-1996 - Application of Safety Instrumented Systems for the Process Industries. Research Triangle Park, NC: ISA, 1996.

2. Goble, William M. *Control Systems: Safety, Evaluation, and Reliability.* Research Triangle Park, NC: ISA, 1988.

3. Goble, William M., and Eric W. Scharpf. "61508 Overview Report," *Safety Lifecycle Report,* <http://www.exida.com>, February 2001.

4. Gruhn, Paul, and Harry L. Cheddie. *Safety Shutdown Systems: Design, Analysis, and Justification.* Research Triangle Park, NC: ISA, 1998.

5. International Electrotechnical Commission. "Functional Safety of Electric/Electronic/Programmable Electronic Safety-Related Systems," IEC 61508, International Electrotechnical Commission.

6. International Electrotechnical Commission. "Functional Safety: Safety Instrumented Systems for the Process Industry Sector," IEC Draft 61511, International Electrotechnical Commission.

7. Scharpf, Eric W., and William M. Goble. "Implementing IEC 61508 in the Process Industries." Presented at SCS2000, Melbourne, Australia, November 2000.

8. U.K. Health and Safety Executive. "Programmable Electronic Systems in Safety-Related Applications, Part 1: An Introductory Guide." Sheffield, UK: UK Health and Safety Executive, 1987.

CHAPTER 3

Tolerable Risk

Organizations have moral, legal, and financial responsibilities to limit the risks their operations pose. Whether the potential recipients of that risk (receptors) are employees or members of the public, they cannot be exposed to a level of risk greater than that which is morally tolerable. In addition to the risks to people, organizations must also consider the risks posed to the environment, property, and business. In some countries, the law mandates tolerable risk levels; in other countries, such as the United States, tolerable risk is determined by each individual organization. These organizations must balance their legal and moral responsibilities with their responsibility to maintain their financial health. Spending resources on risk reduction projects whose costs are grossly disproportionate to their benefits will harm the organization's competitiveness. A good risk management program will therefore carefully select its risk tolerance criteria and balance these three sometimes-competing responsibilities.

Identifying what level of risk is tolerable within an organization can be challenging. Translating its moral, legal, and financial considerations into a workable measure of risk tolerance is a subjective matter that can become quite political. These often conflicting considerations are shown in figure 3.1. Thus, it is important for an organization's risk tolerance staff to gain the buy-in and participation of senior management in order to avoid later second-guessing and rework.

The best path forward in such initiatives begins with a clear understanding of the basic philosophy of risk. It is similarly helpful to learn how to express such risk in clear and objective ways. Finally, it is useful to be aware of what other countries, companies, and organizations have done in the past to identify and act on their own risk tolerances. With this solid foundation, one can properly identify how much risk is too much risk and specify systems to achieve the appropriate level of risk reduction for the project in question.

Inherently, the process of selecting a safety integrity level (SIL) requires decision criteria that convert the estimate of process risk into the required risk reduction, or SIL. Many different methods for selecting an SIL are currently in general use. Some of these methods explicitly use quantitative risk decision criteria. Others use qualitative tools, such as risk graphs, consequence tables, and risk matrices, that tend to obscure the risk criteria on which they are based. Although such methods purport to be qualitative, their end result is quantitative, that is, the SIL. As such, a skilled risk analyst can reverse-engineer those qualitative tools to determine the quantitative criteria behind the tool in question.

Figure 3.1 Risk Management Responsibilities

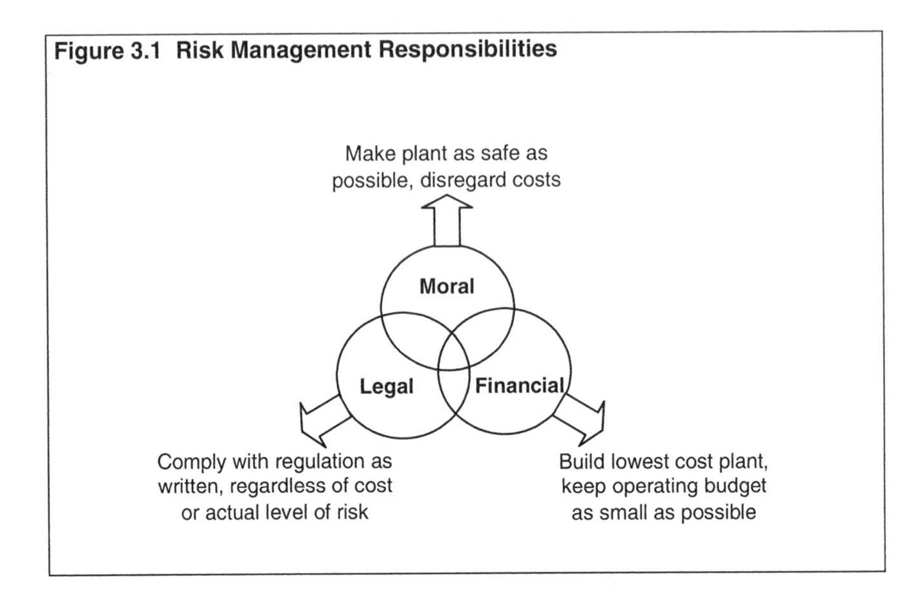

The criteria that help an organization select a risk-reducing SIL will typically help it meet its legal and moral responsibilities, but are generally inadequate for meeting the organization's financial responsibilities. Financial responsibility concerns require organizations to install the safety instrumented system (SIS) with the highest net present value (NPV). Systems with negative NPV should only be installed to meet legal and moral requirements. Including a financial basis in risk reduction planning will enable you to incorporate more risk reduction than using a human and environmental risk-tolerance basis alone. Moreover, using the financial basis will make it easier to justify risk reduction projects to nontechnical managers.

3.1 Philosophical and Political Basis of Risk Tolerance

Many philosophical and political premises can be used to determine whether risks are tolerable or intolerable. The premise any particular organization uses will depend on its specific needs and circumstances. Often, these premises are in competition. Four of the most frequently used premises are the following:

- *Expected Utility Maximization* – This principle is the foundation of cost-benefit analysis and the framework that makes the most sense for industry, although it is rarely used. The difficulty of assigning a monetary value to intangible benefits, such as saved lives, prevents its more widespread use.
- *Rawlsian Approach (Maximize Well-being of Worst-off)* – This approach is often referred to as "social justice" or "wealth redistribution" and

is frequently used by government policy makers. Although this method has an altruistic appeal for some, its economic implications are huge, making it a challenging approach to argue in today's competitive global market economy.

- *Paretian Approach (Make No One Worse Off)* – Although this approach is not entirely practical, a variation of it has found a great deal of favor in both industry and government. When the underlying criterion is relaxed to "making no one *considerably* worse off" and combined with the expected utility maximization approach, this premise is much more appealing. As such, it has formed the basis for many corporate and government risk policies.
- *Nietzschean Elitism (Seek Benefits for Those Who Can Attain the Greatest Benefit)* – While this theory is undemocratic, it is not wholly absent from corporate or even government policy. Producers using this model are maximizing their current output, without any regard for future consequences or who may be impacted. This model poses substantial risks because it diverges from current Western social norms. Thus, it tends to understate the financial loss impacts that result when an organization narrows its definition of risk so acutely. More importantly, adopting this approach would open an organization up to legal intervention should significant harm come to those it has ignored in its initial risk analysis.

All of these theories have been used to varying degrees by government and industry. Most will not work in the long run because technology continually changes, and the information accessible to the general public about technology and industry performance continues to become more available. The philosophy most often used in industry is a combination of Expected Utility Maximization with the Paretian Approach. This hybrid could be expressed as, "maximize the expected utility of your investment, but do not expose anyone, whether your workers or neighbors, to an excessive increase in risk." This method is crystallized by the U.K. Health and Safety Executive's "Tolerability of Risk" framework, which was included in its *As Low As Reasonably Practicable* (ALARP) guideline. The basic components of this ALARP principle are presented in figure 3.2.

The ALARP principle states that there is a level of risk that is intolerable, sometimes called the *de manifestus risk level.* Above this level risks cannot be justified on any grounds. Below this intolerable region is the ALARP region where risks can be undertaken only if a suitable benefit can be achieved. In the ALARP region, risks are only tolerable if risk reduction is impracticable or if the cost of risk reduction is greatly outweighed by the benefit of the risk reduction that is gained. Below the ALARP region and the *de minimus risk level* is the "broadly acceptable"

Figure 3.2 The ALARP Principle

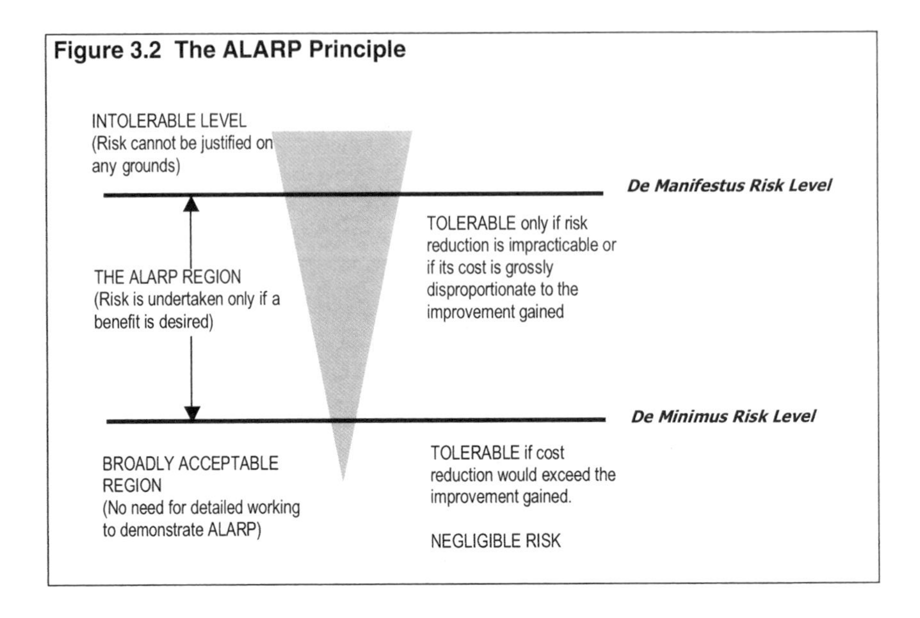

region where the risks are so low that no consideration of them is warranted, and detailed work is not needed to demonstrate ALARP because the risk is negligible. In addition, in this "broadly acceptable" region, risk is so low that no risk reduction is likely to be cost-effective, so a cost-benefit analysis of risk reduction is typically not undertaken.

This ALARP principle can be used with a range of different numeric risk levels defining the boundaries of the three ALARP regions. See section 3.5 for information on the range of boundaries that can be used in different situations.

3.2 Measuring Tolerable Risks (Revealed Values)

When the public is asked about risk in questionnaires, it is usually only able to provide relatively superficial responses. A 1983 study by V. T. Covello demonstrates specifically why questionnaires about risk tolerance tend to be ineffective. It also provides some insight into the reasons qualitative risk assessments tend to be ineffective in general. According to Covello,

- People typically respond to survey questions with the first thing that comes to mind and then become committed to that answer.
- People typically provide an answer to any question posed even though they may have no opinion, do not understand the question, or hold inconsistent beliefs.
- Survey responses are influenced by the order of questions, whether the question is open or closed, supplementary information, interviewer prompting, and how the question is posed.

Clearly, relying on straight questions and answers to determine what risk level most people find tolerable is ineffective. It is far more effective to determine acceptable risk by the indirect method of revealed values. Revealed values can be found by analyzing the types of risks that people currently tolerate, and then quantifying the actual risk those threats pose. In doing so, the researcher is "revealing" the responder's already existing values. This pragmatic approach has been used effectively in several governments' risk criteria, including those of the United Kingdom and Hong Kong.

However, by focusing only on the raw revealed values results alone, without comparing them to other related risks, it is possible to miss important aspects of people's perception and acceptance of risk. For example, important subjective components of risk can be identified when there are significant differences between the revealed value measurement and other measurements of objective risk. These subjective components often relate to the control people feel they have in a given situation and their belief that they can exercise that control more effectively than others. Thus, although air travel causes half as many fatalities as automobile travel in per-hour terms, fear of flying is far greater than fear of riding in a car (1999 World Almanac). Since safety systems must also function in a society where risk is perceived subjectively, they must be designed with these considerations in mind.

3.3 Risk Tolerance Decisions Based on Financial Guidelines

Performing an explicit analysis of the costs and benefits of a risk reduction project, although it is rarely done, is essential to determining the amount of risk reduction that an organization can justify. As we have seen, the ALARP principle states that there is a zone in which process risk should be reduced if reasonably possible. In practice, although not explicitly stated, the loss prevention community has interpreted *reasonable* to mean cost-effective. The best way to determine if a risk reduction project is cost-effective is to calculate the ratio of benefits to costs of a project on a financial basis. If the ratio is greater than one, then the project is cost-effective.

Although this process may seem simple, its application presents pitfalls, the largest of which is determining the benefits of risk reduction. The benefits of a risk reduction project are generally the sum of the decrease in the probability of the following harmful outcomes:

- property damage
- business interruption
- environmental contamination

- injuries to workers and neighbors
- fatalities to workers and neighbors

Calculating a financial benefit for the first two items is not too difficult. However, calculating the benefit of decreases in fatalities, injuries, and environmental damage in financial terms requires that difficult and subtle trade-offs be made. Many organizations simply refuse to perform this type of analysis.

Several methods for determining the value of a saved life have been devised, with varying effectiveness. One method, which is often used to estimate the value of life in tort litigation, is the summation of potential future earnings. A person's annual income is tallied from the date of loss through to their expected retirement with some adjustments for expected raises and inflation. This method has some problematic implications. First, it assumes that the lives of persons with higher-paying jobs are "worth more" than those of low-paid workers. Second, it assumes that retired persons and people who choose not to work, for whatever reason, have no value at all. These social implications and the ability of litigators of wrongful-death torts to routinely obtain sums far larger than a summation of future earnings have rendered this method impractical.

Another approach that has been used to determine the statistical value of a life saved is to determine society's "willingness to pay" for risk reduction by analyzing revealed values. The U.K. Health and Safety Executive has used this method with some success in making its public policy decisions. The following example illustrates how "willingness to pay" can be determined:

> *A person is standing in the middle of a city block and needs to go directly across the street. He can either go to the crosswalk where crossing is safest, but this would require him to expend an additional 30 seconds. Or, he can cross the street where he is and accept the risk of being struck by a vehicle. If he chooses to cross the street where he is he demonstrates that he is not "willing to pay" 30 seconds of his time, which can be roughly valued, for the decrease in risk of collision, which can also be quantitatively determined.*

Although this method has won more acceptance than the summation-of-future-earnings method, especially for making government policy decisions, it has its own drawbacks. A willingness to pay can only be applied to risks that are voluntary because willingness to pay is measured in terms of voluntary risks. As shown earlier, risks that are not voluntary are not as easily tolerated. Extending this line of thought one could argue that it is impossible to determine willingness to pay for an involuntary risk.

The problem with both the willingness-to-pay and summation-of-future-earnings techniques is that they try to provide an ex-ante (before-the-fact) valuation. Most statisticians and engineers agree that analyzing actual operating data (which is by definition ex-post) always yields more accurate information than a predictive model (ex-ante). It would therefore seem natural to estimate the statistical value of a life using historical data for wrongful-death suits. This information is valuable for several reasons:

- It does not represent an attempt to "value" life; it is simply a statistical analysis of the liability imposed on negligence that contributes to a wrongful death.
- It is inherently expressed in financial terms.
- It is based on a broad sampling of social values.
- It is based on exposure to an involuntary risk.

Using this ex-post basis to determine the benefit of preventing a fatality demonstrates that liability stemming from wrongful death is much higher than is implied by either summing future earnings or "willingness to pay." Since wrongful-death cases often have a significant punitive component because of the plaintiff's emotions over the death, the true liability arises both from the objective result of the fatality and the subjective penalty imposed on the defendant's "willful disregard" or "despite best efforts" behavior leading up to the fatality.

3.4 Expressions of Risk

Risk can be expressed in several ways. The measure used to express a risk will depend on two factors:

- Who is being exposed to the risk? Individual employees, individual members of the public, larger segments of society, the environment?
- What is the nature of the risk? Fatality, injury, temporary or permanent environmental damage, financial loss?

The expressions most commonly used for risk to employees' safety are individual risk, societal risk, and geographic risk. Other expressions of risk such as "risk integral" and "expected loss" are also common, especially for financial analyses of risk.

Individual risk is the frequency by which an individual may be expected to sustain a given level of harm from the realization of specified hazards. Individual risk is typically measured as a probability of fatality per year. For process plants, this level of risk is typically determined for the maximally exposed individual.

Societal risk is the relationship between the frequency of a certain level of harm and the number of people suffering from it in a given population when that specified harm is realized. Societal risk is typically shown by plotting the number of fatalities versus the cumulative frequency of events on a log-log chart, commonly referred to as an F-N curve, or Farmer Curve. A typical F-N Curve is shown in figure 3.3.

Figure 3.3 Typical F-N Curve

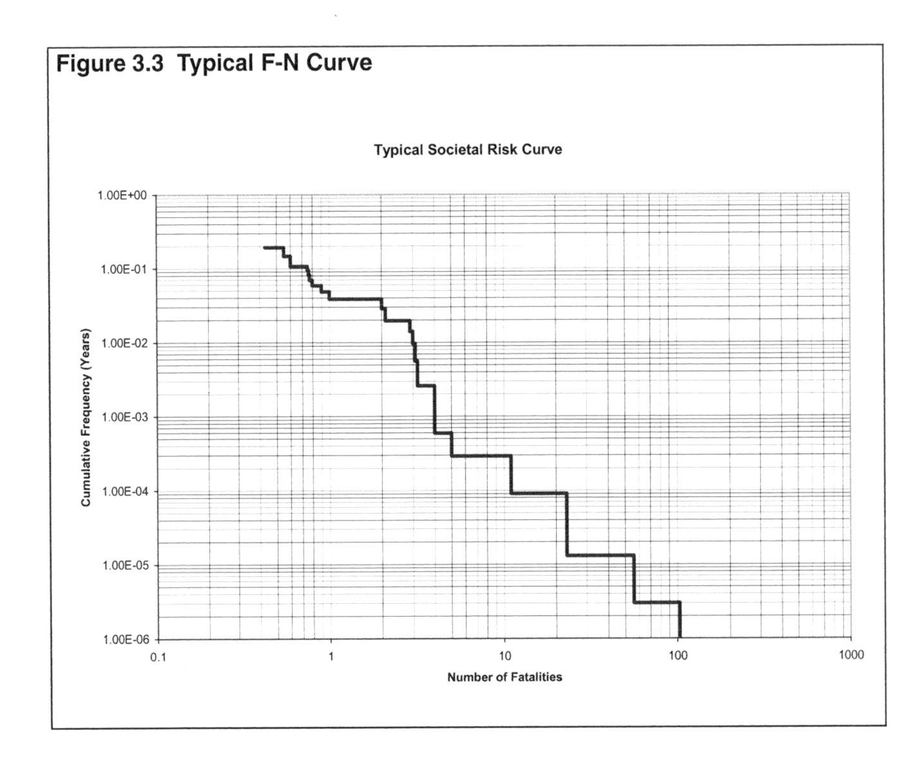

Geographic risk is a measure of the probability that an event will occur in a specific geographic location. Geographic risks are typically shown by drawing constant risk isopleths on a process plot plan, as shown in figure 3.4. Thus, the probability that the designated harmful outcome will occur is the same at each position along the isopleth curve and is generally comparable in the areas of the plot plan in between two isopleths.

A *risk integral* is a summation of risk as expressed by the product of consequence and frequency. The integral is summed over all of the potential unwanted events that can occur and is expressed in equation 3.1. When calculating the risk integral for loss of life, the consequence that is of concern and thus the units of the integral are fatalities.

$$RI = \sum_{i=1}^{n} C_i F_i \qquad (3.1)$$

where:

RI	=	The risk integral
N	=	The number of hazardous events
C	=	The consequence of the event (in terms of fatalities for loss-of-life calculation)
F	=	The frequency of the event

Expected value is a form of risk integral in which the consequences of the unwanted events are expressed on a uniform financial basis. It is often used in calculations to predict insurance loss.

Figure 3.4 Typical Geographical Risk Contours

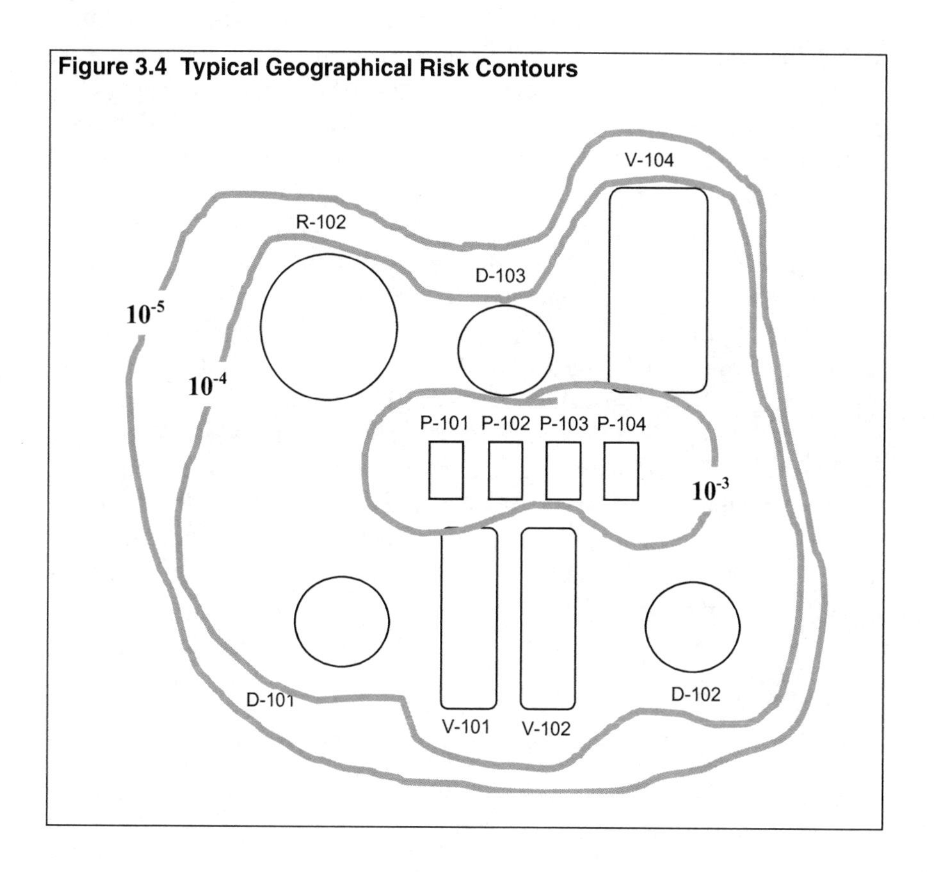

All of these risk measurement criteria have their strengths and weaknesses. Individual risk criteria are probably the most common measures, and are used for everything from facility siting to providing regulatory oversight. The weakness of individual risk criteria is that they do not weigh multiple fatality incidents any more strongly than an isolated incident that causes only a single fatality. Geographic risk criteria are extremely useful for purposes such as facility siting, but they have found little value for risk reduction design engineering. As with individual risk criteria, geographic risk criteria do not weigh multiple fatality incidents more strongly than single fatality incidents.

Societal risk criteria are most widely accepted for tasks such as providing regulatory approval and high-level management oversight of process plants. Societal risk criteria are concerned not only with risks posed to a specific individual, but also risks posed to society in general. This gives them the flexibility to weigh high-consequence events as being less tolerable than low-consequence events. This concept, called "risk aversion," states that given two sets of events that will cause the same number of fatalities, the set of events that causes fewer fatalities *per event* is more acceptable. Ball and Floyd (1998) dispute the claim that there is any social value that justifies "risk aversion"; nevertheless, this concept has been embodied in government criteria, specifically in the Netherlands. Although societal risk provides a more comprehensive picture of risk than individual or geographic risk measures, it is difficult to apply to risk reduction design engineering because it is multi-dimensional. That is, risk is shown as a curve on a two-dimensional chart. Although this visual representation is good for regulatory and management decision-making, it is not straightforward enough for design engineering applications.

Risk integrals have been successfully used in industry and government regulation for a long time. Expected value is probably the oldest quantitative measure of risk and is used extensively in the insurance industry. The premium that an insurance company charges is simply the expected value of loss for the property being covered, with a profit margin added on. Risk integrals are only now gaining acceptance in design engineering as a way to measure risk. They offer several advantages over other methods of measuring risk:

- The risk is expressed as a single one-dimensional value, making it ideal for the design process.
- The risk considers the impact of multiple-fatality events.
- Diverse risks can be expressed on a uniform financial basis, which is essential for cost-benefit analysis.

Because of these advantages, the risk integrals of probable loss of life (PLL) and expected value are ideal for risk reduction design engineering. By using both of these risk measures, organizations can create decision guidelines that meet the requirements of ALARP. Note that PLL can be used in the context of a single event or over a fixed period of exposure or total time. Sometimes this context is not specified so you must take care to apply it appropriately. Another related term used in certain cases is "fatal accident rate" or FAR. It is generally the same as PLL but sometimes has an implied time base of 10^8 hours or about 1,200 workers' full careers. Loss-of-life integrals and expected value can be used to satisfy these needs as follows:

1. Determine a *de manifestus* level of tolerable individual risk (the maximum level of tolerable risk) for both voluntary and involuntary risks by analyzing revealed values.
2. Use this tolerable level of individual risk as an anchor point for determining the tolerable frequency of each unwanted event based on the PLL calculated in the consequence analysis, using the formula listed in equation 3.2. (A derivation of equation 3.2 can be found in appendix A.) This is demonstrated in example 3.1.

$$F_{TOL} = \frac{F_{IND}}{PLL^{\alpha}} \tag{3.2}$$

where:

F_{TOL}	=	The tolerable frequency of a specific event
F_{IND}	=	The tolerable frequency of fatality of an individual (individual risk)
PLL	=	Probable loss of life for a specific event
α	=	Risk aversion factor used to weight high-consequence events more heavily

Thus, an event that has a low PLL can be tolerated more often than an event with a high PLL since it is less likely to cause a fatality in a given number of occurrences.

3. Once it has been established that the frequency of the unwanted event is below the tolerable level just described, the benefits and costs of potential risk reduction projects are calculated using equation 3.3. If the benefit-to-cost ratio is greater than one, the project should be implemented as demonstrated in example 3.2. This corresponds to the middle ALARP region of figure 3.2, where the risk is reduced based on the practicality of being able to reduce it:

$$B - CRatio = \frac{F_{No-SIS}EV_{No-SIS} - F_{SIS}EV_{SIS}}{Cost_{SIS} + Cost_{NT}} \tag{3.3}$$

where:

B–C Ratio	=	The ratio of benefits to costs
$F_{no\text{-}SIS}$	=	The frequency of the unwanted event without an SIS
F_{SIS}	=	The frequency of the unwanted event with an SIS
$EV_{No\text{-}SIS}$	=	The total expected value of loss of the event without an SIS

EV_{SIS}	=	The total expected value of loss of the event with an SIS
$Cost_{SIS}$	=	The total life-cycle cost of the SIS (annualized)
$Cost_{NT}$	=	The cost incurred due to nuisance trips (annualized)

Example 3.1

Problem: A process plant has an individual risk criterion of 2.5×10^{-4} per year. An SIS is being considered to prevent an explosion of the process vessel at the plant. Calculations have shown that the probable loss of life due to the explosion would be 1.75 persons. Based on the probable loss of life and individual risk criterion, what is the tolerable frequency of the explosion event?

Solution: The tolerable frequency of the event is calculated by dividing the individual risk criterion by the expected number of fatalities, or probable loss of life (PLL). As shown in equation 3.2,

$$F_{TOL} = 2.5 \times 10^{-4} / 1.75 = 1.54 \times 10^{-4} \text{ per year}$$

Example 3.2

Problem: An SIS is being installed to prevent a fire that will cost the company $1,650,000. The frequency prior to application of the SIS is once in 1,000 years. After the SIS is installed, it is expected that the frequency will drop to once in 500,000 years. The annualized cost of the SIS is $1,500, and the cost of nuisance trips is negligible. What is the benefit-to-cost ratio of the SIS project?

Solution: Benefit-to-cost ratio can be determined using equation 3.3. The benefit, or the numerator of the equation, is the decrease in frequency multiplied by the expected value of loss that the SIS is protecting against.

Benefit = [(1 / 1,000) × $1,650,000] – [(1 / 500,000) × $1,650,000] = $1,646.70 per year

The cost is the sum of the annualized cost of the SIS and the costs of the nuisance trips, which were stated as negligible.

Cost = $1,500 + $0 = $1,500

And thus the benefit-to-cost ratio is:

B/C Ratio = $1,646.70 / $1,500 = 1.10

A benefit-to-cost ratio of 1.10 means that for every $1 of investment the process owner can expect $1.10 in returns, making the project modestly cost-effective. Note that the margin of error in such calculations is typically much greater than 10% so engineering judgment must be used to make the final decision.

3.5 Benchmarking Risk Acceptance

Once a philosophy and framework for making risk-reduction engineering decisions have been established, numerical criteria need to be selected to ensure that the decision process is systematic and uniform. Typically, an organization may set up separate risk-tolerance criteria for the financial, environmental, and human categories of potential harm. Equivalently, these categories can be lumped into a single variable as a combined risk integral. In either case, it is important to ensure that any comparison to external criteria is performed on an equivalent basis wherever possible. This is best done by using industry and government criteria as benchmarks.

3.5.1 Government and Regulatory Benchmarks

Several nations have developed tolerable-risk criteria. These benchmarks are primarily used for land-planning studies for installations near high-population densities. These criteria either take the form of individual risk criteria, societal risk criteria, or a combination of the two. By far, the government agency that has done the most work in defining and applying the concept of tolerable risk is the Health and Safety Executive of the United Kingdom. Other countries that have government-enforced risk management guidelines include the Netherlands, Hong Kong, Singapore, various states of Australia, and Switzerland. These countries, whose risk tolerance criteria are shown in table 3.1, have largely similar targets, with the exception of the Netherlands. The majority of nations, though, do not have government risk tolerance criteria; risk tolerance criteria are left up to the individual process owners.

Table 3.1 Government Tolerable-Risk Criteria Summary

	UK	Hong Kong	Netherlands	Australia (New South Wales)
Individual Risk de Minimus (Worker)	1×10^{-5}	Not Used	Not Used	Not Used
Individual Risk de Minimus (Public)	1×10^{-6}	Not Used	1×10^{-8}	Not Used
Individual Risk de Manifestus (Worker)	1×10^{-3}	Not Used	Not Used	Not Used
Individual Risk de Manifestus (Public)	1×10^{-4}	1×10^{-5}	1×10^{-6}	1×10^{-6}
Societal Risk Anchor	10 persons at 1×10^{-4}	10 persons at 1×10^{-4}	10 persons at 1×10^{-5}	Not Used
Societal Risk Aversion Index	–1	–1	–2	Not Used

Note: All individual risk values in this table represent annual individual risk of fatality.

The U.K. was the first to develop tolerable-risk criteria, and many of the countries that have risk criteria have largely based theirs on work done in the U.K. The U.K. Atomic Energy Authority began tolerable-risk guidance work in the 1970s when it determined tolerable risks for nuclear installations. This working group developed the Farmer Curve for societal risk (see figure 3.3). Several other U.K. agencies also developed risk criteria, which culminated in the "Tolerability of Risk" framework, of which the ALARP principle is a part.

The Tolerability of Risk guidance quantified levels of tolerable risk. The *de manifestus* level, or maximum tolerable level, of risk for a worker under any circumstances is set at 10^{-3}. This level was chosen because it is approximately the average level of risk faced by any significant group of employees. It is also the same order of magnitude as the risk of death for all employees, regardless of occupation, from all causes. The *de manifestus* level of risk for the public is set at 10^{-4}, which is one order of magnitude less than the level faced by workers. Although this number is somewhat arbitrary, it is on the same order as being killed in a road accident. *De minimus* levels of risk, below which no further risk reduction is required, were set at two orders of magnitude less likely than the *de manifestus* levels of risk for both workers and the public.

The Dutch chose a different route for determining their tolerable risk level. They decided that their tolerable level of risk should be 1 percent of "everyday" risks. Everyday risks were set at the average death rate of the Dutch demographic that had the smallest death rate, that is, young girls. Using this demographic and the 1 percent criterion, a public individual risk level of 10^{-6} per year was developed for *de manifestus* risk.

Hong Kong developed risk tolerance guidelines based on concern over hazardous installations at Tsing Yi Island. The Hong Kong regulators considered the work of the Health and Safety Executive when developing their own criteria, but the guidelines they created were slightly different because of the scope of coverage, which was intended for new housing projects. The Hong Kong guidelines only defined a *de manifestus* level of risk for the general population of 10^{-5}. This is one order of magnitude more conservative than the guidelines created by the U.K. The state of New South Wales in Australia, much like Hong Kong, only sets a *de manifestus* level of risk for the public. They have chosen a value of 10^{-6}, the same as Holland.

Many other countries also regulate risk using quantitative tolerable risk guidelines. These include other states in Australia, Singapore, and Switzerland. Before an organization develops risk tolerance criteria for an application in a specific country it is critical that it consult someone with a thorough understanding of that country's laws and regulations.

The United States is specifically opposed to setting tolerable risk guidelines. Attempts at creating formal decision criteria, sometimes referred to as "bright lines," have always failed. A 1997 report by the Presidential

Commission on Risk Assessment and Risk Management (P/C Commission, 1997) makes the following statement:

> *A strict "bright line" approach to decision making is vulnerable to misapplication since it cannot explicitly reflect uncertainty about risks, population within variation in susceptibility, community preferences and values, or economic considerations – all of which are legitimate components of any credible risk management process.*

Thus, it is unlikely that the United States will adopt tolerable risk criteria any time in the foreseeable future. International standards that attempt to stipulate risk tolerance criteria will also be rejected. Although no government criteria have been imposed on industry in the United States, it still retains an exemplary safety record. The reason for this is partly the flexibility afforded U.S. business to apply capital where it will produce the most benefit and partly the unrestricted ability of the free market to determine third-party liability costs. This free-market action makes most tolerable risk guidelines moot and allows sound decisions about risk reduction projects to be made strictly on a cost-benefit analysis basis.

3.5.2 Industry Benchmarks

Industry has shown a great deal of initiative in defining tolerable risk levels and incorporating them into its internal standard practices. This trend is increasing and is strongest among international corporations, which must meet a number of standards and regulations across the globe. Although internal quantitative tolerable risk guidelines are used more frequently, such criteria are often very confidential.

In 2000, Marszal developed a survey of industry tolerable risk data based on confidential interviews and data accumulated from discussions and publicly submitted reports ("Notes of Confidential Survey and Literature Search"). The variety of industries and company sizes included represented a substantial cross section of safety instrumented system users. A summary of the data found in the survey is shown in table 3.2.

Table 3.2 Summary of Corporate Tolerable Risk Criteria

	High Range	Low Range
De Minimus Individual Risk (Worker)	10^{-5}	10^{-9}
De Manifestus Individual Risk	10^{-3}	10^{-6}
Individual SIF Individual Risk Target	10^{-3}	10^{-6}

Of the companies surveyed, the dominant majority use quantitative criteria for risk tolerance decisions. The balance used either qualitative criteria or standards and codes in lieu of numerical risk criteria. The companies using decision analysis based on standards and codes are not conforming with the safety life cycle principle proposed in IEC 61508 and ANSI/ISA-84.01-1996. Companies using quantitative risk tolerance criteria have shown *de minimus* individual risk ranges from 10^{-3} to 10^{-6}, and the corresponding targets used for selecting SILs also vary from 10^{-3} to 10^{-6}.

Another key finding in the Marszal report is a disparity between the tolerable individual risk level and the target level companies used for SIL selection, for companies that use both criteria. The latter is often one order of magnitude more conservative than the former. The reason for the difference is risk analysts' concern over overlapping impact zones. In many situations, different hazards can impact the same location. To account for this factor, analysts have often simply decreased the SIL selection target level by one order of magnitude. By doing so they hope to ensure that even though hazards may overlap, the overall risk from all sources will still be less than the tolerable individual risk level.

3.5.3 Financial Loss Benchmarks

Financial losses are not commonly used as the sole criterion for risk tolerance. The authors are only aware of a handful of companies that do so, and they strictly use ex-post data to determine the value of fatalities and injuries prevented. Marszal analyzed third-party liability data stemming from injuries and fatalities caused by process plant operators ("Notes of Confidential Survey and Literature Search"). The information from this report is shown in table 3.3.

Table 3.3 Summary of Third-Party Liability from Process Plant Incidents

Loss Type	Low Range	High Range
Fatality	$50,000	$120,000,000
Serious Injury	$10,000	$40,000,000
Minor Injury (Claimed Injury)	$10,000	$40,000,000

The data in table 3.3 show that third-party liability awards vary. The amount of the award depends on a number of factors, including the person who was injured, the location of the accident, and the defendant of the suit. The data show that the value of awards in suits against process plant operators are typically higher than those in suits against other types of defendants. It should also be noted that these results are specific to the United States, although not to any particular region. Settlements in other countries are for the most part much lower than those in the United States (unless the defendant in the suit is a U.S.-based corporation).

Recent jury verdicts in the United States show that because of the high price of third-party liability caused by chemical releases, cost-benefit analysis of this potential liability may become the dominating factor in making decisions about tolerable risk. In late 2000, a jury awarded the estate of a deceased maintenance worker at the Phillips Houston Chemical Complex $120 million because of the fatal injury he sustained as the result of an explosion at the plant. In this case, the jury found the plaintiff to be grossly negligent and awarded substantial punitive damages, but even so the real damages assessed in this case were in excess of $10 million. Given these large amounts, more risk reduction will be justified using cost-benefit analysis than using the most stringent tolerable risk criteria.

3.6 Using a Financial Basis for Making Risk Reduction Decisions

The ALARP principle requires that cost-benefit analysis be used to determine if risk reduction projects should be funded when they fall into the ALARP region. By the Health and Safety Executive's own admission, the *de manifestus* and *de minimus* risk levels were set so that most process risks fall into this intermediate region. As such, most risk reduction decisions will require a cost-benefit analysis. Since this is true, cost-benefit analysis should be built into the SIL selection process. Several companies are just beginning to employ this concept, and they are achieving excellent results. They have found that, for the most part, the tolerable risk guidelines they have set on a moral-legal basis are almost never used because the financial aspect of the risk reduction project always justifies a greater amount of risk reduction.

Mudan, Shah, and Myers (1995) prepared a report for the Loss Prevention Symposium that describes this use of financial risk management by describing the relationships between various process hazards. They discovered that risk caused by third-party liability of personnel injury is insignificant in comparison to other losses such as property damage and business interruption. For refineries, property damage losses always dominate, and for upstream refining operations business interruption losses always rule. This study found that making risk reduction engineering decisions based on personal risk levels alone is inadequate because it ignores the major risk to the corporation, which is financial.

3.7 Summary

Organizations have legal and moral obligations to limit the amount of risk posed by the processes they operate and derive benefit from. The challenge of engineering risk reduction equipment such as safety instrumented systems is to identify quantitative targets upon which to base one's designs. Most guidelines in use today are basically attempts to quantify "soft" statements regarding what is morally acceptable, even if those guidelines are codified into law.

We have established that the ALARP principle is a good foundation for justifying a level of tolerable risk. To effectively employ ALARP, we need to create a more precise definition of the level of risks beyond which no projects are acceptable, regardless of the benefit obtained. This is the *de manifestus* risk level. We also need to determine the level of risk below which risk is considered negligible and analysis of the situation is not required. This is the *de minimus* risk level. In between the *de minimus* and *de manifestus* levels of risk is the ALARP region, where risk reduction efforts should be determined on the basis of cost-effectiveness. In addition, we have determined that voluntary and involuntary risks should be measured separately since they are tolerated at different levels. The quantitative levels that organizations should use for these targets are left up to each individual process operator, but they should conform to local regulations and be justifiable in terms of what others in the industry are using.

A two-step approach is generally required for risk tolerance guidelines that fulfills the requirements mentioned in the previous paragraph and that provides the most cost-effective risk reduction measures. First, one must ensure that the risk is reduced to a level below the *de manifestus* level of individual risk, based on the probable loss of life (PLL) or other equivalent criteria for the incident under review. Second, one must then perform a cost-benefit analysis to determine the optimal amount of risk reduction that can be justified financially. This analysis should consider all of the benefits of the risk reduction project, including not only decreased loss of life and injury but also decreased property damage, environmental contamination, and business interruption. All of these should be viewed on a uniform financial basis.

3.8 Exercises

3.1 What does ALARP mean? What are the three levels of risk described by the ALARP principle?

3.2 Describe two ways to determine the financial value of a saved life.

3.3 List two common measures of risk.

3.4 An organization has maximum individual risk criteria of 4.5×10^{-5} per year. A toxic release at the plant is expected to result in 20.5 fatalities. What is the tolerable risk of this event expressed in terms of frequency?

3.5 A major fire, which can result in a probable loss of life of 0.1, is expected to occur at a frequency of 3.9×10^{-4}. The plant where this hazard exists has accepted this risk as tolerable with no further risk reduction. Based on this information, the plant's tolerable individual risk of fatality is greater than what value?

3.6 For what range of benefit-to-cost ratios are risk reduction projects cost-effective?

3.7 Referring to figure 3.4, if a person were to spend his or her entire work life of 2,000 hours per year in between vessel V-101 and V-102, what is the average annual risk to which he or she has been exposed? (Assume that when not at work he or she is exposed to zero risk.)

3.8 A plant is considering installing an automatic fire water system. The fire water system has a probability of failure on demand of PFD = 0.05. The fire that the system would prevent would cause $35 million in damage if the system did not operate, but only $5 million if the system did operate. The fire is expected to occur with a frequency of once in 1,900 years. The annualized cost of the fire water system is $5,930 per year. Is the fire water system a wise investment? Support your answer with the calculation of a benefit-to-cost ratio for the fire water system.

3.9 True or false?: The United States has set maximum individual risk-of-fatality thresholds for 1×10^{-3} per year for workers and 1×10^{-4} for the general public.

3.9 References

1. American Petroleum Institute. "Management of Hazards Associated with Location of Process Plant Buildings," API RP-752. Washington, DC: API, 1995.
2. ANSI/ISA-84.01-1996 - Application of Safety Instrumented Systems for the Process Industries. Research Triangle Park, NC: ISA, 1996.
3. Ball, David J., and Peter J. Floyd. *Societal Risks: A Report Prepared for the Health and Safety Executive*. London: HM Stationery Office, 1998.
4. Center for Chemical Process Safety. *Guidelines for Safe Automation of Chemical Processes*. New York: American Institute of Chemical Engineers, 1993.
5. Covello, V. T. "The Perception of Technological Risks: A Literature Review," *Technological Forecasting and Social Change* 23 (1983): 285-97.
6. Harvey, B. H. (chairman). *First Report of the Advisory Committee on Major Hazards*. London: HM Stationery Office, 1976.
7. Health and Safety Executive. *The Setting of Safety Standards: A Report by an Interdepartmental Group of External Advisors*. London: HM Stationery Office, 1996.
8. International Electrotechnical Commission. IEC draft standard 61511, Part 3, "Guidelines in the Application of Hazard and Risk Analysis." Geneva: IEC, 1999.
9. Lees, F. P. *Loss Prevention for the Process Industries*. London: Butterworth and Heinemann, 1992.
10. Marszal, E. M. "Notes of Confidential Survey and Literature Search on Tolerable Risk Guidelines and Third Party Liability Settlements and Judgments." Columbus, OH: Exida, 2000.
11. Marszal, E. M., B. A. Fuller, and J. N. Shah. "Utilization of Risk Based Criteria for Safety Integrity Level Selection," *Process Safety Progress* 43, no. 34 (2000): 75.
12. Mudan, K. S., J. N. Shah, and P. M. Myers. *Financial Risk Assessment: A Uniform Approach to Manage Liabilities*. New York: American Institute of Chemical Engineers, 1995.
13. Smith, D. J. *Reliability, Maintainability, and Risk*. London: Butterworth and Heinemann, 1993.

CHAPTER 4

Identifying Safety Instrumented Functions

As noted in the safety life cycle described in chapter 2, once the scope of a process is defined, its associated risks and hazards can be identified. This identification leads naturally to the corresponding identification of the safety instrumented function (SIF) that may be required to reduce the process risk to a tolerable level. In this chapter we discuss methods for identifying these required SIFs.

To identify a safety instrumented function properly, you should have a clear understanding of what one is. IEC 61508 defines a safety function as a "function to be implemented by an E/E/PE [Electric/Electronic/Programmable Electronic] safety-related system, other technology safety-related system or external risk reduction facilities, which is intended to achieve or maintain a safe state for the EUC [Equipment Under Control], in respect of a specific hazardous event." A safe state is defined as when the EUC is in a state of "freedom from unacceptable risk." IEC 61511 defines a safety instrumented function as an "E/E/PE safety function with a specified safety integrity level which is necessary to achieve functional safety. A safety instrumented function can be either a safety instrumented protection function or a safety instrumented control function."

Translating all of this into less technical terms, an SIF is the action that a safety system takes to make the overall process safe when confronted with a potential hazard. This action must take place if the process is to achieve a safe state with a probability designated by the function's safety integrity level (SIL).

Thus, it is important to clearly and precisely identify what these safety instrumented functions must accomplish before determining how reliably they must accomplish it. The first thing the function must accomplish is to signal or indicate that a hazard is present and that harm will result unless something is done. The next consideration is to determine what safe state should result from the function for this process. The final issue is what the function must specifically do to achieve that safe state result. Thus, with a clear idea of what constitutes an indication to act, of what the safe state result is, and of what specific action is required, you have identified a safety instrumented function.

This exercise is often not as easy as it sounds, however. For instance, suppose a compressor needs to be shut down if it goes into surge. This shutdown involves stopping the motor, opening the recycle valve, and closing the discharge valve. What constitutes the critical SIF action? If the harm is that the compressor could suffer severe damage and the safe state is full recycle, then the recycle valve opening is the critical function. The answer to what constitutes the critical SIF action depends on the particulars of the system in question. What this means in practice is that control, equipment, and process knowledge are needed to identify the full SIF properly.

Fortunately, as part of the required hazards and risks analysis, this combination of control, equipment, and process expertise should already have addressed these issues in reasonable detail. Ideally, this analysis will provide most of the information needed to put together the complete list of possible safety functions for that process.

The challenge posed by this risk identification and hazards analysis is how to ensure that it is complete without spending millions of hours on the question. The best way to balance cost and completeness is to use a systematic approach that is supported by a diverse team of process and safety experts who are capable of providing several different perspectives on the process. In a systematic approach, the team will be focused on working efficiently and the whole procedure will be streamlined by delegating follow-up outside any group meetings. Although this is by no means easy, it should not be avoided just because it sounds too hard. If at this point the team has not identified an important safety function, it will be difficult to later assign it an SIL and rely on it to prevent a potentially severe accident.

4.1 General Risk Identification and Hazard Analysis

There are several different accepted methods for identifying process risks and analyzing hazards. Most of them take the form of a proactive study called a *process hazard analysis* (PHA). PHA is required in the United States by the OSHA Process Safety Management and EPA Risk Management Plan legislation. The "Seveso Directive" (96/82/EC) places similar requirements on process operators in the European Union. Most methods of PHA consist of a structured brainstorming exercise in which a team of experts systematically reviews sections of a process to identify possible hazards. The PHA team also lists the events that can cause accidents, the potential outcome of these accidents, and the safeguards in place to prevent the accidents. Finally, the PHA recommends other measures the organization should implement to reduce process risk.

Since most processes and systems are rarely completely new, a less formal method for identifying SIFs is to directly incorporate functions identified during previous processes. For instance, many oil refinery processes have been operating since the early 1900s with very few changes. After accidents and near misses occur, good practice compels engineers to determine the root causes and install protective systems that prevent the incidents from reoccurring. This reactive approach to risk reduction is incorporated into the process of designing new units by engineering companies that have long histories designing particular types of process units. The engineering firm begins the design process for each new unit with template drawings that include the risk reduction measures that have been integrated over the life of the process. They take extreme care to take into account any differences between previous applications of the process and the one they are now considering.

A wide variety of PHA methods are used in process plants. The type used depends on the complexity of the process under study, the amount of experience that an organization has with the process, and whether the plant is new or undergoing a review. The most popular PHA method is the *hazards and operability (HAZOP) study.* A HAZOP study uses guide word combinations to help a team of experts identify failure scenarios that can cause process accidents and operability problems. First, the team breaks the entire process into smaller, more manageable sections called "nodes." Nodes are typically chosen by looking at the natural process equipment and function breaks present in the overall system. Similarly, a set of guide words such as "too much pressure," "too little flow," etc., is chosen to support the review. The team then systematically applies the guide word combinations to each part of each node to identify which hazards are present, whether there are any existing safeguards, and if any additional safeguards are needed. HAZOP studies are most effective when the process is complex and unique.

A *checklist study* is a type of PHA in which a team of process experts asks a list of questions that may identify process hazards. This type of analysis is very effective when the process under study is small or when there are many identical or very similar processes. For instance, checklist PHA studies are often used for LPG distribution facilities and chlorine injection processes in municipal water treatment plants. The checklists that are developed for these processes consider scenarios where accidents and near misses have occurred in other similar process plants. Many checklist studies for process units are based on the requirements stated in process-specific standards.

A *what-if study* identifies hazards by posing questions about failures that could occur in the process. To be effective, what-if studies must be conducted by a good facilitator who asks the right questions of an appropriately broad team of process and safety experts. The questions, which the facilitator often develops, are based on expert knowledge of the likely failure modes of the process equipment being used and of the hazards specific to the materials in the process. The what-if study is effective in situations where the process equipment and the hazards of the process materials are well understood and the risks are perceived to be low because the consequences are small. What-if studies are typically used in combination with checklists.

Another systematic PHA method is the *failure modes and effects analysis* (FMEA), which focuses on what causes a failure and the effects the failure has on the system. This procedure first determines a level of detail or resolution for the analysis and then identifies the standard failure modes. Next, the analysis establishes a frequency and consequence severity scheme for the effects of those failures. The results are typically cast in the form of a table that lists the different components, failures, causes, and effects. The results can also be ranked in terms of the risks associated with the system.

Whatever PHA an organization uses, the end result is identifying hazards and/or harmful events. The PHA team's discussions of each hazard will also include the hazard's consequence. If one of the reverse methods is used, the discussions of each harmful consequence will include each event that can lead to that consequence. In either case, the PHA team will gather information on the safeguards in place to prevent an accident. In addition, the PHA team should recommend new safeguards or at least call for further investigation when it feels that the existing safeguards are inadequate.

Selecting safety integrity levels (SILs) begins with developing a list of SIFs that should be analyzed to determine what, if any, SIL is required to reduce the risk of the process to a tolerable level. SIFs can be readily identified from several different sources of information.

4.2 Identification from PHA Reports

As we have mentioned, the PHA report contains important information on the hazards present in the process. A review of the PHA report will yield the SIFs that are currently in place to prevent accidents as well as any additional SIFs that the PHA team has recommended to improve safety. PHA reports often present results in a tabular format. For each hazard identified, several characteristics are described, such as any safeguards used or recommendations made to improve safety. The safeguards column will contain the SIFs currently being used in the process, and the recommendations column will contain the new SIFs that should also be considered. Thus in most cases, the PHA report provides a wealth of vital information for identifying SIFs—an important step in the SIL selection process.

4.2.1 PHA Report Structure

Table 4.1 shows an excerpt from a typical PHA report, in this case a report for a HAZOP study. The report has a tabular format in which each identified hazard is listed in a row, and the characteristics of the hazard are listed in the columns. There are variations in the data reported among the various styles, but most present the same general information. The report shown in table 4.1 contains columns for causes, consequences, safeguards, and recommendations. This report also shows the initial process deviation guide words, in this case, "more pressure," that caused the hazard to be identified.

The main difference between the PHA reports generated from the various methods is the prompt that caused the hazard to be identified. In a HAZOP, the prompt is a guide word combination. In checklist studies, the prompt is a checklist item, while in a what-if study the prompt is the what-if question posed by the facilitator. Regardless of the type of study, the consequence of the hazard, the safeguards that prevent the hazard, and the recommendations for improvement should all be listed, usually in a tabular format.

Table 4.1 Excerpt from Typical HAZOP Report

Debutanizer Column Node: Reboiler Section

Dev	Cause	Consequence	Safeguards	Recommendations
1.0	More Pressure			
1.1	Column steam reboiler pressure control fails, causing excessive heat input	Column overpressure and potential mechanical failure of the vessel and release of its contents	Pressure relief valve, operator intervention to high-pressure alarms, mechanical design of vessel	Install SIF to stop reboiler steam flow upon high column pressure
1.2	Steam reboiler tube leak causes high-pressure steam to enter vessel	Column overpressure and potential mechanical failure of the vessel and release of its contents	Pressure relief valve, operator intervention to high-pressure alarms	See Item 1.1
2.0	Less Flow			
2.1	Low flow through bottoms pump causes pump failure and subsequent seal failure	Pump seal fails and releases flammable material	Low outlet flow Pump Shutdown SIS	Existing safeguards adequate

Note that Dev stands for deviation which refers to the guide word in question as it is applied to the specific section of the specific node under consideration.

4.2.2 Information Required from the PHA Report

The following four types of information will be required when developing a list of safety instrumented functions from a PHA report as part of the SIL selection process:

1. **A description of the safety instrumented function (SIF).** The description of each safety instrumented function, which is either already in place or has been recommended, will be found in the safeguards and recommendations columns of the report, respectively. These reports often note certain items that require further investigation. These items must be followed up and included as additional functions if further investigation results suggest this.
2. **A description of the consequence being prevented**. The hazards that are being prevented can be found in the same row as the safety instrumented function identification in the consequences column.
3. **The initiating events that can cause the consequence.** The initiating events that cause an accident to occur can be found in one of two places. Either the report will contain a causes column, or the cause of the hazard will be found as part of the question that prompted the hazard to be identified. For instance, the cause might be found in a what-if question or a checklist item.
4. **The other safeguards that are available to prevent each initiating event from propagating into the accident.** Safeguards other than the SIF under consideration will be found in the safeguards column. It is important to list not only the non-safety instrumented safeguards but also other SIFs that might perform an action that will prevent the same hazard. These other safeguards are also important in the layer of protection analysis (LOPA) described in chapter 9.

The first step in the SIL selection process is to identify all of the SIFs that need to have their SILs selected. SIFs are generally listed in two locations in a PHA report. Existing safeguards are listed in the safeguards column, and recommended safeguards that do not already exist in the process are listed in the recommendations column.

The PHA report excerpt in table 4.1 shows two SIFs. Scanning the safeguards column, we find a low outlet flow pump shutdown SIF in the third row (2.1) of the report. Since the SIS is listed in the safeguards column, this SIF already exists in the process. Scanning the recommendations column of the PHA report yields one additional SIF. The first row (1.1) contains a recommendation to "Install SIF to stop reboiler steam flow upon high column pressure." Since the SIF is in the recommendations column of the PHA report, this SIF is recommended and does not already exist in the process.

The hazard that is being prevented and its associated consequence can be found in the consequences or description of hazard column of the PHA report. In the PHA report excerpt of table 4.1, the "More pressure" consequence that has been identified is "column overpressure and potential mechanical failure of the vessel and release of contents."

Initiating events are related to the consequence that was determined in the previous step. The initiating events leading to that consequence should be determined for each row that has a consequence that needs a SIF. In a HAZOP report, the initiating events are found in the causes column of the PHA report. In what-if and checklist studies, the initiating events might also be found in the prompts (e.g., checklist items and what-if questions). Multiple initiating events may potentially be present for a single consequence. When this situation occurs, the same consequence will be shown in multiple rows of the PHA report. Each row where the consequence appears will have different causes.

In the PHA report excerpt of table 4.1, the consequence "column overpressure and potential mechanical failure of the vessel and release of its contents" appears in two rows. For this situation, there are two causes or initiating events related to this outcome: (1) column steam reboiler pressure control fails, causing excessive heat input, and (2) steam reboiler tube leak causes high-pressure steam to enter the vessel.

Each individual initiating event will have an associated set of safeguards. The listed safeguards are unique to each initiating event and are not generic to the consequence that is being prevented. The PHA report makes it easier to determine safeguards for each individual initiating event because the protection layers listed in the safeguards column are always associated with an individual initiating event. Each row of the report is dedicated to a single initiating event.

In the PHA report excerpt, there are two initiating events, both of which lead to the same consequence. The initiating event "column steam reboiler pressure control fails, causing excessive heat input" has the safeguards of pressure relief valve, operator intervention, and mechanical design. The initiating event "steam reboiler tube leak causes high-pressure steam to enter vessel" has the safeguards of pressure relief valve and operator intervention. Note that the set of safeguards for the pressure control failure is not the same as the set of safeguards for the reboiler leak scenario.

Reports and documentation from the required process hazards analysis are an excellent source of information on existing and recommended SIFs. Usually this information is clearly presented in one of the forms described earlier. If this is not the case and the report is either unclear or the information is not present, the PHA itself may not satisfy the required government regulations. In that case, the PHA procedure should be reviewed to bring it up to standards. The references listed at the end of this chapter provide more detailed information on how to conduct a proper PHA.

4.3 Identification from Engineering Drawings

Even when the PHA is properly conducted and well documented, a review of the engineering design documents for the process is also recommended to ensure that all SIFs have been identified. Process licensors and detailed design engineering firms will often include SIFs in their basic design package. These functions are usually shown in documents such as piping and instrumentation diagrams (P&IDs) and process flow diagrams. Identifying SIFs based on engineering documents requires an understanding of control engineering, the process under study, and risk analysis. Although SIFs might be included in basic engineering design packages, they are often not differentiated from basic process control functions, which can complicate the process of identifying them.

Even a good PHA often fails to identify hazards that have occurred in the past, for which safeguards have already been added to the process. While these SIFs may not appear in the PHA report, they will be shown on the engineering design documents. Detailed design contractors and process licensors typically incorporate the operating experience for processes into a design package without much fanfare. The process designers often include SISs in new designs to prevent accidents and near misses that have occurred in other similar process designs without specifically noting this in concise report form.

Thus, identifying SIFs from the engineering design package is often difficult to do since they are typically not differentiated from the other control loops in the process. To identify SIFs based on a P&ID representation of a control loop, you must have control engineering expertise so you can clearly understand the function that is being performed. You must also have a knowledge of the process under study to be able to understand why each of the control functions shown on a P&ID is being employed.

When identifying SIFs by reviewing design documentation, layer of protection analysis (LOPA) can be complicated. This stems from the fact that much of the information normally generated by a broad team of experts and provided in a PHA must instead be developed ad hoc by the safety analyst. There is typically no document that describes the hazard being prevented, the consequences if the hazard is realized, or the safeguards that are in place to prevent the hazard. The safety analyst must develop all of this information through his or her expertise in risk analysis and a thorough understanding of the process under study. To do this effectively, expert assistance from either the process design engineers in the detailed design or process licensing firm and/or a risk management consulting firm may be required. In short, the same breadth of expertise and support needed for a good PHA is often needed to provide the equivalent information on the SIFs not noted in the original analysis.

4.4 Summary

The following three steps summarize the procedure for developing the information needed to systematically select SILs for all of the SIFs that exist or have been recommended:

1. **Obtain the design documents.** These documents should include the process hazards analysis (PHA) report, the licensor's piping and instrumentation diagrams (P&IDs), and the detailed design contractor's P&IDs.
2. **Review the PHA report.** Search the safeguards list to identify the SIFs that already exist in the process, and search the recommendations list column for newly recommended SIFs. For each SIF that has been identified, use the PHA report to determine the consequences that the SIF is protecting against, the initiating events that can cause that outcome, and the protection layers that help prevent each initiating event from developing into the unwanted accident.
3. **Review the piping and instrumentation diagrams.** Use control engineering expertise to determine which control loops might be performing safety functions and process expertise to verify that a safety function is being performed. Then use risk analysis expertise to identify the consequence, initiating events, and protection layers. This additional review should add to the list of SIFs that have been generated from the PHA report.

Again, it is extremely important to properly identify all of the potential safety instrumented functions. These SIFs are the actions a system takes to prevent a hazard from causing severe harm to equipment, personnel, or the environment. Performing the process hazards analysis (PHA) and risk identification steps is vital to identify all of the SIFs potentially required for a process. PHAs are systematic studies that use prompts to help a team of process experts identify the potential hazards of a process. Although PHAs are typically required by U.S. and European regulations, conducting a PHA is usually a good business decision since the costly accidents it prevents almost always pay for the effort in the long run. These PHA methods differ mainly in the type of prompting that is given to the PHA team. The more common methods include HAZOP, what-if, and checklist. Although we discussed the basics of the common methods in this chapter, see the sources listed in the reference section for the more in-depth procedures you must follow to conduct these analyses rigorously.

PHA reports provide a wealth of information on the SIFs that exist and have been recommended for a process. The report itself is typically tabular, with each row representing a hazard and the columns representing characteristics of the hazard. Columns usually include the consequences if the hazard is realized, the causes of the hazard, the safeguards in place to prevent the hazard from developing into an accident, and recommendations for improving process safety.

In cases where the PHA report does not identify all of the SIFs that exist in a process, we recommend reviewing the design documentation, including the P&IDs from the process licensor and/or the detailed design contractor. To identify the SIF, you will need control engineering expertise because SIFs are not always marked as such. Determining the balance of information needed to build a LOPA diagram also requires that you have specific process knowledge and risk analysis expertise.

4.5 Exercises

4.1 List two sources of information that can be used to identify SIFs that are present or recommended for a process.

4.2 List two methods that are used to perform process hazard analysis (PHA).

4.3 Identify the SIFs that are required for the process shown in the following PHA report, including initiating event(s), consequence(s), and safeguard(s).

Dev	Consequence	Cause	Safeguards	Recommen-dations
1.0	More Level			
1.1	Acrylonitrile feed tank is overfilled, resulting in a liquid evaporation of the spilled material and downwind exposure to toxic gas	Delivery truck contains more material than the feed tank is capable of containing	Process operator stops material transfer upon receiving a high level alarm	Install SIF to the transfer pump when a high feed tank level is detected

4.4 What is an SIF?

4.6 References

1. ANSI/ISA-91.00.01-2001-Identification of Emergency Shutdown Systems and Controls That Are Critical to Maintaining Safety in Process Industries. Research Triangle Park, NC: ISA, 2001.

2. Center for Chemical Process Safety. *Guidelines for Hazard Evaluation Techniques.* Washington, DC: American Institute of Chemical Engineers, Center for Chemical Process Safety, 1992.

3. Lees, F. P. *Loss Prevention for the Process Industries*. London: Butterworth and Heinemann, 1992.

4. Occupational Safety and Health Administration. *Process Safety Management (Pamphlet OSHA 3132)*. Washington, DC: U.S. Department of Labor, Occupational Safety and Health Administration, 1993.

CHAPTER 5

Rules of Probability

Many of the risk analysis techniques and tools organizations use to systematically select safety integrity levels (SILs) are based on a foundation of probability and statistics. This chapter explains some basic probability and statistical concepts for determining both the likelihood and the consequence of unwanted events.

We will discuss how to assign probability to a single event by examining either the physical characteristics of the event or by using experimental or historical data. We will also discuss the ways in which events are logically related to each other and how those relationships can be expressed in mathematical terms. For instance, an equipment item can either be working or in a failed state, so the probabilities of the working and failed events are complementary and mutually exclusive. When events are logically related—such as A and B must occur simultaneously to create an unwanted event—you can determine the probability of the combined events by using probability math. This chapter discusses how to use probability multiplication to determine the probability of events that are logically AND'ed and probability addition to determine the probability of events that are logically OR'ed.

This chapter also discusses basic fault tree analysis and how it can be used to model the probability of end results or "top events" that are based on complicated combinations of causal events related by often complex logical structures. We also describe the symbols that represent different types of basic events and the logical relationships of events, called *gates*, along with the basic methods of calculation. Although most of the basic concepts of probability are straightforward, the complexity of the applications of those concepts can make the work challenging.

5.1 Assigning Probability to an Event

Probability is a mathematical expression of the chance that something will occur. It is used to determine the odds of the outcome of an event occurring when one or more possibilities for an outcome exist and when the overall basis for that outcome is completely random. The probability of an event's occurrence can be described as the number of times that it will occur divided by the number of trials. This relationship is shown in equation 5.1. Probability P is a real number that always lies between zero and one. A probability of one means an event will always occur, and a probability of zero means an event will never occur.

$$P = \frac{n}{N} = \frac{Occurrences}{Trials} \tag{5.1}$$

The probability of an event is determined by one of two methods, either by analyzing the physical properties and characteristics surrounding the event or by empirically analyzing historical or experimental data. When considering physical properties, the analyst must consider the number of potential outcomes, the degree of randomness in the system, and how each of the outcomes is related to each other. Consider a typical pair of dice used in games of chance. Since each die has six sides, the number of possible outcomes when one of the dice is rolled is six. If the die is fair, then each side has an equal chance of coming up, and the probability of any particular side coming up is one divided by six, or P = 1/6.

Another way that probability is determined is by analyzing the outcome of empirical data such as experimental trials or historical records. As with analysis of physical properties, the probability is determined using equation 5.1. In this case, the total number of trials is the number of records in the data set, and the number of occurrences is the number of records where the outcome of interest occurred.

5.2 Types of Events and Event Combinations

To perform reliability engineering and frequency analysis, we first need to define the generic events that occur in process plants in terms of probability theory. The type of event, whether independent or dependent of random chance, will determine the appropriate model for calculating a single event's probability. The type of event combination, whether complementary, mutually exclusive, or non-mutually exclusive, will determine the probability equations we must use when calculating the overall probability of event combinations. Thus, the event type and the combination type together specify the model and probability equations needed for any given problem.

5.2.1 Independent Events

Independent events have outcomes that are not related in any way. When two events are independent, the outcome of one event does not affect the probability of another subsequent event. For example, the fact that one coin toss resulted in heads does not impact the outcome of the next coin toss. Examples of independent events include tossing coins, spins of the roulette wheel, and dice throws.

Reliability analysts normally consider the failure of one component of a complex system to be an independent event. For example, if a pressure relief valve and a non-return check valve are used in the same process pipe to prevent an accident, it is usually valid to assume that the failures of these two devices are independent of each other.

In some cases, a single common cause failure can cause multiple pieces of equipment to fail simultaneously. In these cases, analysts must give special consideration to the calculation of their reliability. Although analysts typically assume that component failures are independent, the analyst must carefully review the process situation to ensure that the events are truly independent enough to make that assumption. The suitable degree of independence is, unfortunately, often a judgment call based on engineering experience. However, since the assumption of independence or common cause susceptibility can dramatically affect the results of a probability model, this assessment must be made with extreme care. For instance, in our example of the non-return check valve and pressure relief valve, the process fluid may be highly viscous or corrosive, so common cause failure of both components is a real possibility.

5.2.2 Complementary Events

A pair of *complementary events* is a situation in which only one of two outcomes is possible. If one event occurs, then by definition, the other event cannot. For instance, when a rounded-edge coin is tossed, the result must be either heads or tails. There is no way that the toss can result in heads and tails, or neither. Therefore, heads and tails form a pair of complementary events.

When performing reliability engineering, it is often assumed that a device's failure state and operational state are complementary events: the device is always either operational or failed. It does seem obvious that the device cannot simultaneously be working and failed. However, there are numerous situations where this is not rigorously true. Consider a relief valve whose set pressure is 500 psig. The pressure in the protected vessel rises above 500 psig all the way to 565 psig before the pressure is relieved, but the vessel is relieved before it ruptures. Has the relief valve operated or failed? Although one could argue that it neither operated nor failed, in practice one must define *operation* and *failure* more precisely to prevent such ambiguities. For example, a relief valve is said to be working if it safely relieves the system pressure at +/– 5% of its set pressure. For our purposes, we will assume precise definitions exist in the real world, that a device has either failed or is working, and that these events are therefore complementary.

Since one event of a set of complementary events is always true, the probability that one event out of the set is true is 100% or 1.0. Imagine that there are two complementary events, A and A*. The probability that one of the events has occurred is 1, as shown in the equation 5.2:

$$P_A + P_{A^*} = 1 \tag{5.2}$$

From this, we can easily derive the basic formula for determining the probability of one of two complementary events given that we know the other:

$$P_{A*} = 1 - P_A \quad (5.3)$$

Example 5.1

Problem: The probability of the successful operation of a check valve for the next year is 0.89. What is the probability of failure for the same year?

Solution: The probability of failure, or $P_{FAILURE}$, will be the complement of the probability of success, or $P_{SUCCESS}$. Using equation 5.3, we see that

$$P_{FAILURE} = 1 - P_{SUCCESS}$$

$$P_{FAILURE} = 1 - 0.89 = 0.11$$

Again, the calculation involved here is trivial when performed in the isolated application. The challenge comes in ensuring that each of these simple equations is used properly in the long sequence of calculations required to analyze complex systems.

5.2.3 Mutually Exclusive Events

A mutually exclusive set of events exists when only one of a set of outcomes is possible. No two mutually exclusive outcomes can occur simultaneously. For example, when one die is tossed the outcomes 1, 2, 3, 4, 5, or 6 are mutually exclusive. The roll of the die cannot result in 2 and 3 at the same time. Thus, the complementary events of the last section are a subset of *mutually exclusive events* where there are only two possible outcomes in the set.

Other than the mutually exclusive (and complementary) assumption that a piece of equipment has a "failed" or an "operational" state, mutually exclusive event sets are rarely used in reliability engineering. Failures of equipment are usually assumed to be non-mutually exclusive. This means that more than one piece of equipment can be in the failed state simultaneously. For instance, a transmitter that senses a process upset and a valve that is closed to protect the process against the effects of that upset might both be in the failed state at the same time. In this case, failure of the valve and of the transmitter are not mutually exclusive.

5.3 Combining Event Probabilities

We can determine the probability of event combinations from the individual events that make up the combination. The event combination probability is a function of the following:

- The probability of the individual events
- The logic that relates the individual events to each other
- The types of event sets (for instance, independent and mutually exclusive)

Two basic logical combinations are crucial for reliability engineering: the logical AND and the logical OR. A logical AND is an operation in which the result of a statement is TRUE if and only if all of the input statements are TRUE. When independent events are combined using a logical AND, we can determine the probability of the outcome by using probability multiplication of the input statement probabilities. A logical OR is an operation in which the result of a statement is TRUE if any one or more of the input statements is TRUE. When independent events are combined using a logical OR, we can determine the probability of the outcome by using probability addition of the input statement probabilities.

5.3.1 Probability Multiplication

As mentioned, we determine the probability of the combination of independent events that are logically AND'ed by using probability multiplication. When performing a probability multiplication, simply multiply all of the probabilities of the input statements P_i, as shown in equation 5.4, to get the combined probability P:

$$P = \prod_{i=1}^{n} P_i \tag{5.4}$$

For example, if there are two independent input statements, A and B, then the probability of A AND B occurring is P_A times P_B.

5.3.2 Probability Addition

We can determine the probability of the combination of independent events that are logically OR'ed by using probability addition. When performing a probability addition, we calculate the probability of the outcome differently depending on how the events are related, that is, whether the events are mutually exclusive or can occur simultaneously. Although this difference is easy to forget, it can be extremely important in many calculations.

Example 5.2

Problem: A limit switch and a solenoid valve must both work in order for a control system to function. In the next year, the probability of the limit switch operating successfully is 0.8, and the probability that the solenoid valve will operate successfully is 0.95. What is the probability of success for the system assuming that the switch and valve can be treated independently?

Solution: The probability of success is the probability that both the limit switch AND the solenoid valve successfully operate. Using equation 5.4 for this situation yields:

$$P_{SUCCESS} = P_{LIMIT\ SWITCH} \times P_{SOLENOID}$$

$$P_{SUCCESS} = 0.8 \times 0.95 = 0.76$$

If the set of input statements is mutually exclusive, probability addition will simply be a matter of adding all of the probabilities of the input statements P_i, as shown in equation 5.5, to arrive at the total probability P. If, for instance, there are two independent input statements, A and B, and they are mutually exclusive, then the probability of A OR B occurring is P_A plus P_B.

$$P = \sum_{i=1}^{n} P_i \tag{5.5}$$

Example 5.3

Problem: A fair die is rolled. What is the probability that either a 4 or a 5 is rolled?

Solution: The probability of rolling a 4 is 1 out of 6, or 1/6. The same applies to the probability of rolling a 5. Since the various outcomes that are possible for the roll of a die are mutually exclusive, we use the mutually exclusive form of probability addition, as shown in equation 5.5:

$$P_{(4\ OR\ 5)} = P_4 + P_5$$

$$P_{(4\ OR\ 5)} = 1/6 + 1/6 = 2/6$$

When the events are not mutually exclusive, the situation becomes more complex. Consider the Venn diagram of event A and event B in figure 5.1, where A and B are not mutually exclusive.

The probability of each independent event occurring is represented by the area of each event's circle. The probability that either event occurs is a union of these two sets and would be represented by the area of both circles together. Since the events are not mutually exclusive, the two sets overlap. The overlap represents the situation where both events occur

Figure 5.1 Venn Diagram of Non-Mutually Exclusive Events

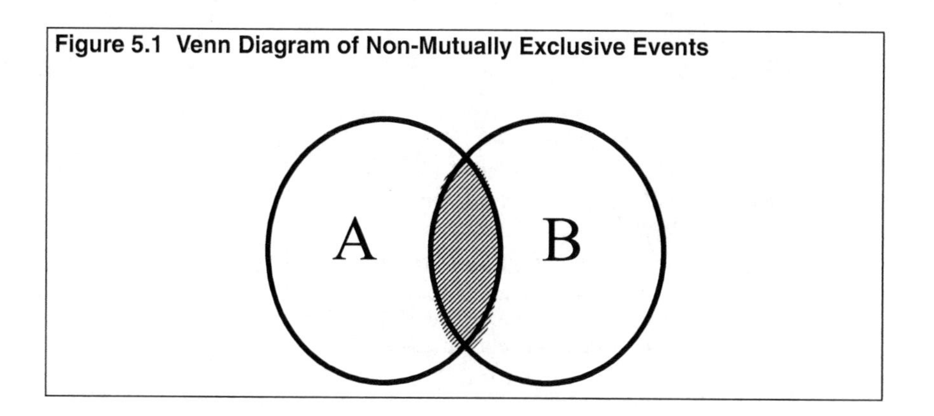

simultaneously. If we calculate the probability of A OR B using simple addition, the overlap section will be counted twice. We therefore must subtract the area of the overlap. The area of the overlap is calculated using probability multiplication since it represents the probability that A AND B occur. After making this key adjustment, the final equation that we use for the probability addition of non-mutually exclusive events is as follows:

$$P_{AorB} = P_A + P_B - P_{AandB} = P_A + P_B - (P_A \times P_B) \quad (5.6)$$

Equation 5.6 is only valid when there are two events. There are many situations in which we need to calculate the probability addition of more than two events. As such, we must have a general equation for the probability addition of multiple events. Equation 5.7 is the general form of the probability addition equation for any number of non-mutually exclusive events. In general, the probability of multiple events is one minus the product of one minus each of the individual probabilities.

$$P = 1 - \prod_{i=1}^{n} (1 - P_i) \quad (5.7)$$

Consider an example in which an outcome depends on three events, A, B, and C, and these events are non-mutually exclusive. Using the general equation for probability addition of non-mutually exclusive events, the probability that A or B or C will occur is as follows:

$$P = 1 - (1 - P_A)(1 - P_B)(1 - P_C) \quad (5.8)$$

Example 5.4

Problem: A sack contains 100 marbles. All of the marbles are either spotted or striped, and all of the marbles are either green or red. Sixty percent of the marbles are spotted; ninety percent of the marbles are green. If a marble is randomly selected from the bag, what is the probability that the marble will be spotted OR green?

Solution: The events SPOTTED and GREEN are not mutually exclusive because it is possible to withdraw an object that is both spotted and green. Thus, the non-mutually exclusive form of probability addition is used. For two events, equation 5.6 can be applied as follows:

$$P_{SPOTTED\ OR\ GREEN} = P_{SPOTTED} + P_{GREEN} - (P_{SPOTTED} \times P_{GREEN})$$

$$P_{SPOTTED\ OR\ GREEN} = 0.6 + 0.9 - (0.6 \times 0.9) = 0.96$$

Note that the mutually exclusive form of the probability addition equation would yield an answer of 1.5.This would then violate the definition of probability, which cannot exceed the upper limit of 1.

Equation 5.7 can also be applied in this situation, yielding an identical result:

$$P_{SPOTTED\ OR\ GREEN} = 1 - (1 - P_{SPOTTED}) \times (1 - P_{GREEN})$$

$$P_{SPOTTED\ OR\ GREEN} = 1 - (1 - 0.6) \times (1 - 0.9) = 0.96$$

Example 5.5

Problem: Consider a system composed of a transmitter, controller, and valve. The probability of failure, over the next five-year period, for each of the components is as follows:

$P_{f,transmitter} = 0.15$

$P_{f,controller} = 0.008$

$P_{f,valve} = 0.19$

Over the next five-year interval, what is the probability of success of this system?

Solution: For this system to operate successfully the transmitter AND the controller AND the valve must operate successfully. Since the events are logically AND'ed, we must use probability multiplication.

While the problem asks for the probability of success, it describes the reliability of the individual components in terms of probability of failure. To determine the probability of success, we assume that failure and success are complementary events, and use equation 5.3 to calculate the probability of failure.

$P_{success,transmitter} = 1 - 0.15 = 0.85$

$P_{success,controller} = 1 - 0.008 = 0.992$

$P_{success,valve} = 1 - 0.19 = 0.81$

Example 5.5 continued...

The probabilities of success are then multiplied:

$P_{Success} = P_{success,transmitter} \times P_{success,controller} \times P_{success,valve}$

$P_{Success} = 0.85 \times 0.992 \times 0.81 = 0.683$

Note that the overall probability of success is lower than the probability of success of the lowest component. This will always be true.

A second way of looking at this same problem is to treat it in the reverse way. For successful operation of the system, the transmitter OR the controller OR the valve CANNOT fail. The probability of any one of these failing is calculated from equation 5.7 as follows:

$P_{FAIL\ T\ OR\ C\ OR\ V} = 1 - (1 - 0.15) \times (1 - 0.008) \times (1 - 0.19) = 0.317$

Then, since the problem asks for the probability of success, we take the complement, following equation 5.3, to arrive at:

$P_{Success} = 1 - P_{FAIL} = 1 - 0.317 = 0.683$

which is the same answer.

Thus, as long as the logic is sound, the path to the solution does not matter.

5.4 Fault Tree Analysis

Fault tree analysis is a top-down approach for describing the failures of complex systems. A fault tree analysis begins with the "top event," which is the result of a number of basic events that contribute to, or initiate, the system failure. The logic of a fault tree is displayed by the symbols that represent the basic events and gates that logically relate those events. Each of the common fault tree symbols represents a type of event or a logical relationship.

Fault tree analysis can be a very powerful tool for analyzing the frequency or probability of an accident or failure of a piece of equipment when simple probability math alone cannot determine the outcome. Fault tree analysis not only represents the way events are logically related; it can also quantify the probability of those events. Additional analysis allows us to determine various parameters such as the importance, uncertainty, and sensitivity of a system.

The following discussion of fault tree analysis provides a very basic description with just enough detail to make it possible to do some safety integrity level selection tasks. You can quantify the fault tree by using one of three methods: (1) minimal cut set analysis, (2) gate-by-gate analysis, (3) Monte Carlo simulation. This section describes the gate-by-gate method, since it is the most straightforward and suitable for the SIL application at hand. More information on minimal cut set analysis and

Monte Carlo simulation can be found in *Loss Prevention for the Process Industries* by F. H. Lees.

5.4.1 Fault Tree Symbols

Figure 5.2 shows some of the common symbols used to describe the logical relationships in a fault tree model. While many symbols have been defined for fault tree analysis, only a small fraction are in common use. The most common symbol is the rectangle, which is used for a top or intermediate-level event. The *top event* is the fault, harm, or unwanted accident under study. *Intermediate events* are the logical result of other, more basic events, and occur as a precursor to the top event.

Two types of primary events are shown in figure 5.2, the basic event and the undeveloped event. Basic or primary events are the events at the bottom of the fault tree. No other events or logic gates provide input to a basic event. The basic event, represented by a circle, needs no further causal breakdown because its failure rate or probability of failure can be easily quantified or understood. An undeveloped event, which is represented by a diamond, is one that has not been analyzed to determine the more basic events and logic that comprise it. Often, the undeveloped event is left as such because its probability is qualitatively determined to be so low that no further consideration is warranted.

Logic gates are used to relate basic events to intermediate events and in turn to the top event, both qualitatively and quantitatively. The two gates that are most commonly used are the OR gate and the AND gate.

A number of additional symbols, not described here, are defined in standards and texts on fault tree analysis. These symbols define more complex logical relationships, such as NAND (not and), NOR (not or), and voting relationships. Also, symbols for managing large fault trees that require multiple pages have been defined. As stated previously, these additional symbols are typically not needed for the fairly simple fault trees developed during the process of selecting SILs.

5.4.2 AND Gates

The AND gate performs a logical AND operation as one would expect. When the fault tree is used qualitatively, the AND gate means that the gate's output is true if all of the inputs are true. When using the fault tree quantitatively, the output of an AND gate is determined by using probability multiplication, assuming that the input events are independent.

5.4.3 OR Gates

The OR gate performs a logical OR operation. When the fault tree is used qualitatively, this means that the gate's output is true if any one or more of the inputs is true. When using the fault tree quantitatively, the output of an OR gate is determined by using probability addition. As noted previously, the form of the probability addition equation used will

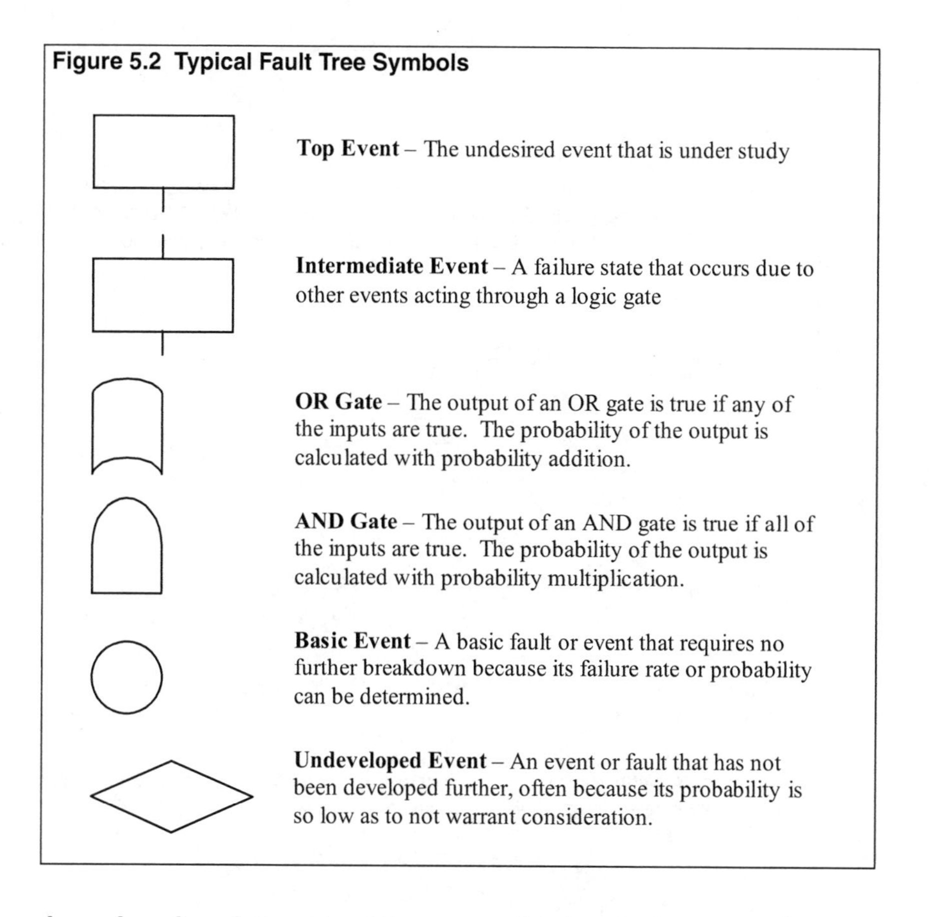

Figure 5.2 Typical Fault Tree Symbols

depend on the relationship of the events, that is, whether or not they are mutually exclusive. In the field of risk and reliability analysis, it is generally correct to assume that the failure of pieces of equipment is non-mutually exclusive. Typically, more than one equipment item in a group can be failed simultaneously.

5.4.4 Combining AND and OR Gates

To present a slightly more complicated problem and to illustrate the importance of independence relative to common cause we will now look at example 5.6 under a slightly different set of assumptions. In this case, we assume that half the time feed A fails it is because of a fault common to both A and B, which also causes feed B to fail.

We can see the dramatic increase in total system failure rate that results from a common cause. In addition, we can see that translating the logical structure of the fault tree into equations can quickly become quite complicated. The important thing to keep in mind when solving these more involved problems is to consider each step one at a time to be sure it properly represents the situation you are modeling.

Example 5.6

Problem: A power supply system consists of two separate feeds. Each feed is supplied by a separate power company and is independently wired (i.e., the power feeds are independent). The fault tree shown below illustrates the situation. The probability of failure during a one-year period of feed A is 0.015, while 0.108 is the probability of failure of feed B. What is the probability that the power supply system will fail to operate over the course of one year?

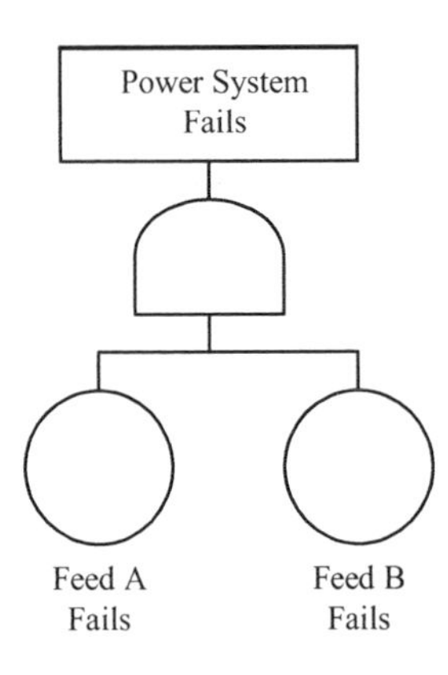

Solution: Solve the fault tree using a gate-by-gate analysis. The output of the AND gate is determined using probability multiplication.

$$P_{\text{SYSTEM FAILS}} = P_{\text{FEED A FAILS}} \times P_{\text{FEED B FAILS}}$$

$$P_{\text{SYSTEM FAILS}} = 0.015 \times 0.108 = 0.0016$$

Example 5.7

Problem: A water coolant supply system consists of two pumps; one is electrically driven and the other is steam driven. Both pumps are continuously operating and together must supply an adequate amount of cooling water. The following fault tree illustrates the situation. The probability of pump A's failure, over a one-year period, is 0.02, and 0.03 is the probability of pump B failing. What is the probability that the cooling water system will fail to operate over the course of one year?

Example 5.7 continued...

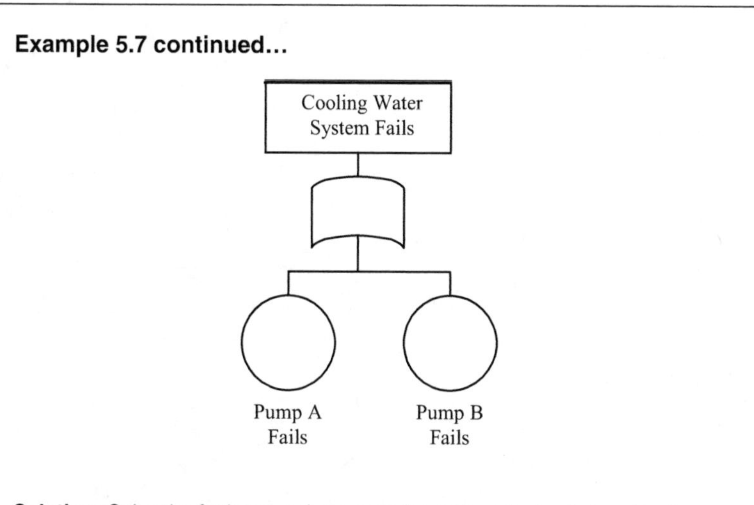

Solution: Solve the fault tree using a gate-by-gate analysis. Determine the output of the OR gate using the non-mutually exclusive form of probability addition.

$$P_{SYSTEM\ FAILS} = P_{FEED\ A\ FAILS} + P_{FEED\ B\ FAILS} - (P_{FEED\ A\ FAILS} \times P_{FEED\ B\ FAILS})$$

$$P_{SYSTEM\ FAILS} = 0.02 + 0.03 - (0.02 \times 0.03) = 0.0494$$

or equivalently from equation 5.7:

$$P_{SYSTEM\ FAILS} = 1-(1-P_{FEED\ A\ FAILS}) \times (1-P_{FEED\ B\ FAILS})$$

$$P_{SYSTEM\ FAILS} = 1-(1-0.02) \times (1-0.03) = 0.0494$$

Consider what would occur if you used the simpler but incorrect form of the probability equation. This common mistake produces the result 0.05 instead of 0.0494. This only represents an error of 1%, and the direction of the error is conservative, which means that the failure rate determined incorrectly is higher than the error rate that is calculated using the proper form of the equation. Because the error is often small and conservative, risk analysts often use the simpler, mutually exclusive form of the probability addition equation, regardless of the ways the inputs are related. This shortcut is explained in more detail in section 5.6.

Example 5.8

Problem: A power supply system consists of two separate feeds. Both feeds are supplied by the same power company but are independently wired from the local transformer. The following fault tree shown illustrates the situation. The probability of failure during a one-year period of feed A is 0.015, while 0.108 is the probability of failure of feed B. However, because the feeds come from the same power company, half of the failures of feed A have the same cause as the failures of feed B. What is the probability that the power supply system will fail to operate over the course of a year?

Example 5.8 continued...

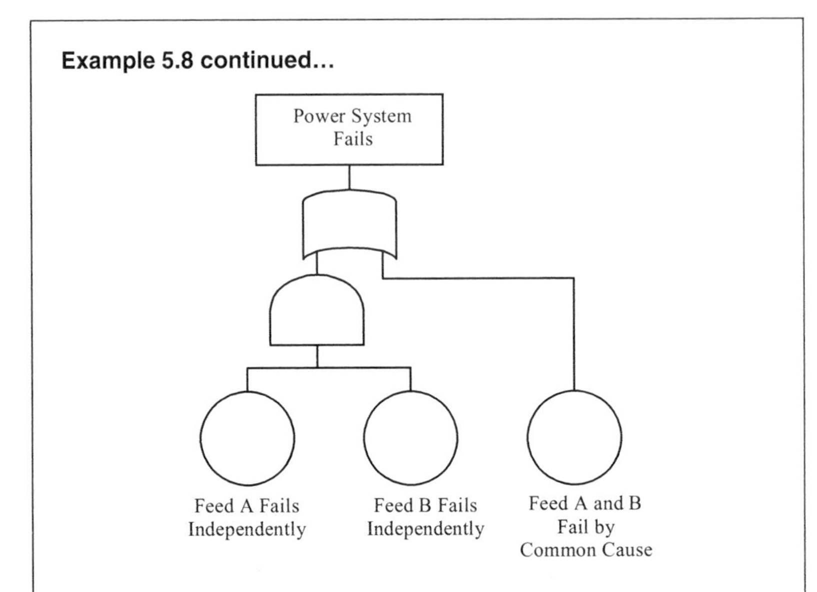

Solution: Solve the fault tree using a gate-by-gate analysis. First determine the basic event probabilities.

The probability that feed A fails independently is half of its total failure rate, so

$$P_{\text{A FAILS INDEPENDENTLY}} = P_{\text{FEED A FAILS}} \times 0.5 = 0.015 \times 0.5 = 0.0075$$

The probability that A and B fail by common cause is then the other half of A's failure rate since the common cause failure and the independent failure are mutually exclusive.

$$P_{\text{A and B FAIL BY COMMON CAUSE}} = P_{\text{FEED A FAILS}} - P_{\text{A FAILS INDEPENDENTLY}}$$

$$= 0.015 - 0.0075$$

$$= 0.0075$$

The remaining probability that B fails independently is then the difference between B's total failure rate and the common cause rate since the common cause and independent failures are mutually exclusive.

$$P_{\text{B FAILS INDEPENDENTLY}} = P_{\text{FEED B FAILS TOTAL}} - P_{\text{FEED A and B FAIL BY COMMON CAUSE}}$$

$$= 0.108 - 0.0075 = 0.1005$$

As before, we determine the output of the AND gate using probability multiplication.

$$P_{\text{A and B FAIL INDEPENDENTLY}} = P_{\text{FEED A FAILS}} \times P_{\text{FEED B FAILS}}$$

$$P_{\text{A and B FAIL INDEPENDENTLY}} = 0.0075 \times 0.1005 = 0.000754$$

Since the inputs to the OR gate are mutually exclusive, we simply add them as follows to get the final result:

$$P_{\text{SYSTEM FAILS}} = P_{\text{A and B FAIL INDEPENDENTLY}} + P_{\text{A and B FAIL BY COMMON CAUSE}}$$

$$P_{\text{SYSTEM FAILS}} = 0.000754 + 0.0075 = 0.00825$$

Note that the addition of a 50% common cause increases the probability of total system failure by a factor of 5 compared to the result from example 5.6.

5.5 Failure Rate and Probability

Failures of industrial equipment are usually reported in terms of a *failure rate* or a *mean time to failure*. These failures are not often presented in terms of probability of failure on demand for a number of reasons. First, the probability of failure on demand is only of interest when the piece of equipment is a protection layer whose failure will not be detected until it needs to act. When the failure initiates a chain of events that leads to an accident or a nuisance shutdown, failure rate is a more appropriate measurement. Second, the probability of failure of a piece of equipment that does not initiate an accident, or a shutdown, will depend on how often the equipment is tested. Test frequency becomes quite important when one is designing a system to meet a specific SIL requirement. We discuss this further in chapter 10 in the context of assigning SILs.

5.5.1 Determining Failure Rates

To quantify equipment failure rates you need historical information on how events that cause or propagate an accident have occurred in the past. The best source of this data is the process plant's record of previous failures and equipment maintenance. This information is best because the failure rate applies to the actual conditions under which the process equipment is being used. Information gathered from another plant or from general industry databases might not be representative because the equipment is maintained differently or used in more severe process conditions. Unfortunately, this plant-specific historical reliability data is often not available in readily usable form.

Most operating companies keep maintenance and repair records for process equipment, which can also help determine relevant failure rates. Many national laws and regulations require that companies keep such records. For instance, in the United States maintenance records for safety-related equipment items are required by the mechanical integrity clauses of the OSHA Process Safety Management (PSM) regulation and the EPA Risk Management Plan (RMP) regulation. In addition, industry consensus standards such as ANSI/ISA-84.01-1996 require that safety instrumented systems be periodically tested. The records from these tests are another valuable source of data on how well the components in question can be expected to perform.

Industry average data is also available from consortiums of operating companies and engineering societies such as ISA and AIChE. They collect such information for a range of operating conditions and categorize it into generic equipment types. It is unlikely to find failure rates for a specific model of instrument in a generic industry database, but specific applications are often included there. This data can provide some support and background context when direct information is insufficient and the situation calls for engineering judgment.

The quantification process will inevitably require you to use some expert judgment. This is so because the failure rates obtained from several different sources may vary substantially, and the failure rates in one particular database may apply over a large variety of equipment and thus be expressed as a range of values. Applying this data to a specific situation will require the analyst to compare and consider the differences between the equipment installed in the plant and the equipment described in the database or other source. Since this expected failure rate can significantly affect the SIL selected, one must take extreme care to seek as much information as possible to confirm any given rate.

Given the number of times a piece of equipment has failed and the overall mission time of that equipment, it is relatively easy to estimate the failure rate provided two key assumptions are made. First, you must assume that the failure rate is constant so the failure is equally likely to occur at any point in time. This assumption is widely used in reliability engineering, and is accurate for most equipment throughout most of its useful life. In this case, we ignore the higher incidence of failure typically found when equipment is initially installed and when it nears the end of its useful life. Second, you must assume that the number of failures that has been recorded constitutes an accurate sample. When drawn in the form of a histogram, the time to failure based on multiple test results should form a bell curve in which the center of the curve indicates the mean time to failure (MTTF). The greater the number of data points in the histogram, the more accurately one can predict the MTTF (and failure rate). If the number of failures in the data set is small (for instance, less than 10), it will be difficult to accurately predict the MTTF. In this case, you should infer the failure rate by assuming a failure rate distribution and selecting a confidence level. More information on inferring failures rates using distributions and confidence levels can be found in *Reliability, Maintainability, and Risk* by David Smith.

Assuming that the failure rate is constant and that there is sufficient data, then the failure rate λ is simply the number of failures divided by the total mission time of the equipment. This is shown in equation 5.9.

$$\lambda = \frac{k}{T} = \frac{Number\ of\ Failures}{Mission\ Time} \tag{5.9}$$

5.5.2 Failure Rate and Mean Time to Failure

Another common measure of the reliability of process equipment is the MTTF, or *mean time to failure*, which is related to that equipment's failure rate. When the failure rate is constant, the mean time to failure is simply the inverse of failure rate, or one over the failure rate λ as shown in equation 5.10.

Example 5.9

Problem: A plant has 157 relief valves, each of which is subject to an annual function test. The record of these function tests shows that over a five-year period, there have been three instances of valves failing to pass the function test. What is the failure rate for this plant's relief valves?

Solution: The failure rate is the number of failures divided by the mission time of the equipment, as shown in equation 5.9. The number of failures is three (i.e., k=3), as given in the problem statement.

The amount of time in operation, in hours, is five years, multiplied by 157 valves:

$$T = 5 \text{ years} \times 157 \text{ valves} \times (8760 \text{ hours/year}) = 6.87 \text{ million hours}$$

Thus, the failure rate is:

$$\lambda = 3 \text{ failures} / 6.87 \text{ million hours} = 0.44 \text{ failures per million hours}$$

or equivalently 0.0038 failures per year.

Failures per million hours and failures per year are the units most commonly used to describe failure rate but FIT – failures per billion hours (10^9 hrs) – is also sometimes used.

$$MTTF = \frac{1}{\lambda} \tag{5.10}$$

Example 5.10

Problem: What is the MTTF of the relief valves analyzed in example 5.9?

Solution: Example 5.9 assumed that the failure rate for this situation is constant, so equation 5.10 is valid. Thus the MTTF is:

$$\text{MTTF} = 1 / (0.44 \text{ per million hours}) = 2.27 \text{ million hours}$$

or equivalently 259 years.

5.5.3 Industry Average Data Sources

As mentioned earlier, when plant specific data is not available, you can obtain useful information from generic compilations of failure rate data. Many data compilations are publicly available in which the data is segregated based on industry type and device type. For instance, separate data sets are available for the offshore industry and nuclear industry even though they use much of the same type of equipment. The failure rates published for a specific piece of equipment can span several orders of magnitude if the variability of the data is high. Many published data banks show three different failure rates: a low rate representing devices that are stressed less than average, a mean rate, and a high rate for devices that are in services more severe than average. The failure rates in most published databases are presented in terms of "failures per million hours," which requires the analyst to convert the failure rate into the units appropriate for the study at hand. A fair amount of expert judgment must be used when applying generic data to a specific application. The analyst must weigh whether the relevant situation is more or less severe than average and select the proper failure rate accordingly.

Some of the most popular sources of failure rate data are shown here. This list represents just a sampling of the available databases; many others, both public and private, also exist.

- *Guidelines for Process Equipment Reliability Data* – This reference is published by the Center for Chemical Process Safety (CCPS) of the American Institute of Chemical Engineers (AIChE). It is a compilation of individual studies based on onshore bulk and specialty chemical production operations.
- *Offshore Reliability Data* – This book, published by Det Norske Veritas, is a compilation of data from several companies operating offshore platforms in the North Sea. The book is usually referred to as *OREDA* followed by the year of publication. For instance *OREDA 97* is the most recent version. Although the data is limited to offshore applications, it is widely used in other industries due to the wide range of devices included and the large amount of data.
- *Nonelectronic Parts Reliability Data* – The Reliability Analysis Center (RAC), which is an information analysis center for the United States Department of Defense, publishes this reference. It is also referenced by its acronym (NPRD) and the year of publication. The most recent version of this publicly available book is *NPRD-95*. The data in the book includes summary reliability data and detailed descriptions of part types. It also contains reference data for a wide variety of parts, including general mechanical components, as well as process, instrumentation, and control equipment.
- *Reliability, Maintainability, and Risk* – This publication, written by David Smith and published by Butterworth-Heinemann, is a reliability-engineering textbook that contains extensive appendices of failure rate data for equipment ranging from large process equipment to small electronic devices.
- *Loss Prevention for the Process Industries* – This three-volume collection is the definitive reference on loss prevention engineering. It contains several data tables of equipment reliability information, mainly for onshore and offshore process plant equipment. This resource also contains human reliability data and other data tables unavailable elsewhere, such as ignition probabilities useful in flammable spill calculations.

5.5.4 Probability of Failure on Demand

When performing the frequency analysis that accompanies SIL selection, various representations of failure rate are required. For instance, with layer of protection analysis (LOPA), initiating events must be expressed in terms of frequency (or rate) while the protection layer and mitigating events need to be in the form of probabilities. Thus, the end result of the analysis, the likelihood of unwanted events, will be expressed in terms of frequency. LOPA is discussed in more detail in chapter 9.

Since probabilities are required to determine the likelihood of protection layer failure and mitigating events, they have to be determined based on failure rates. When determining the probability that a process safeguard has failed, we are concerned about its probability of failure on demand (PFD). As one would expect, the PFD of a device depends on the device's failure rate. It also depends on the failure mode of the equipment (whether that mode is safe or dangerous) and the proof test interval (how often the device is tested).

The failure mode of a device is the symptom, condition, or way in which a device fails. A failure mode might be identified as a loss of function, premature functioning (functioning without demand), an out-of-tolerance condition, or simply a physical characteristic such as a leak (incipient failure mode) observed during inspection. In some cases, the failure rate that is listed in a database is divided into those failures that will cause a failure on demand, or dangerous failures, and those that will cause a nuisance trip, or safe failures. These databases provide an overall failure rate that includes both modes as well as a distribution between safe and dangerous failures. This presentation is common in failure rate databases for electronic components.

Most databases list the failure rates for each mode of failure. For process equipment items, many failure modes may exist. For instance, in the *CCPS Reliability Guidelines*, the failure modes listed for pilot-operated relief valves include all of the following items (note that it is important to select the proper mode of failure when quantifying a LOPA diagram):

- Seat leakage
- Fails to open
- Spurious operation
- Fails to open on demand
- Inter-stage leakage

The probability of failure on demand for a device depends on its frequency of testing and repair. An untested device's PFD gets larger as time increases since failures tend to be non-correcting. For a constant failure rate, the relationship between failure rate and test interval is exponential, as described by equation 5.11:

$$PFD_{Max} = 1 - e^{-\lambda t} \qquad (5.11)$$

where,

PFD_{Max} = the maximum failure probability, or the one at the end of the time interval in question

λ = the failure rate

t = the mission time, or time between complete function tests

Example 5.11

Problem: A relief valve has a failure rate of 1.68 failures per million hours. The failure mode for that rate is "fails to open on demand." This valve is checked and tested annually. What is the maximum probability of failure on demand of the relief valve?

Solution: The failure rate is converted into failures per year:

$$\lambda = 1.68 \frac{Failures}{10^6\ Hours} \times 8760 \frac{Hours}{Year} = 0.015 \frac{Failures}{Year}$$

The PFD_{Max} is calculated using equation 5.11:

$$PFD_{Max} = 1 - e^{-0.015 \times 1} = 0.015$$

In section 5.6 we will show how to take advantage of the fact that the exponential function gave a PFD_{max} that is essentially equal to the failure rate λ.

It is important to note that the PFD calculated in equation 5.11 is the maximum failure probability over the entire test interval or mission time. While using the maximum failure probability is a safe and conservative practice, it is more correct to use an average failure probability. The demand for a safeguard to operate can occur at any time during the test interval with an equal probability (assuming constant failure rates). Since this is true, it can be unnecessarily conservative to always apply the failure probability that reflects the full test interval.

We calculate the average failure probability over a time interval by integrating the failure probability function, as shown in equation 5.11, over the test time interval, and then dividing by the test time interval, as shown in equation 5.12.

$$PFD_{avg} = \frac{\int_{t=0}^{T} \left(1 - e^{-\lambda t}\right) dt}{T} \qquad (5.12)$$

While this kind of integration is possible it becomes somewhat cumbersome for most applications, and is not often used. Instead, average failure rates are calculated using a simplification of the failure probability equation. Using this simplification, we can gain a good estimate of the average failure rate through equation 5.13. We justify the use of the simplified equation for failure probability in section 5.6. A derivation of equation 5.13 can be found in appendix A.

$$PFD_{avg} = \frac{\lambda t}{2} \tag{5.13}$$

Example 5.12

Problem: What is the average failure probability of the relief valve described in example 5.11?

Solution: The PFD_{avg} is calculated using equation 5.13.

$$PFD_{avg} = (0.015 \times 1) / 2 = 0.0075$$

5.6 Simplifications and Approximations

In the previous sections of this chapter we presented a number of equations for calculating the probability and failure rate of events. Although these equations are important for determining the risk posed by process hazards, in practice some are replaced with more compact approximations. The compactness of the approximations makes it possible to perform calculations more rapidly and to manipulate them more easily.

IMPORTANT: It is important to remember that these approximations are primarily used to speed the calculation process. It is critical that the analyst evaluate the specific situation to ensure that the approximations described here will provide reasonable and sufficiently accurate results before applying them in practice.

5.6.1 Simplified Probability Equation

The probability that an event will take place during a given time period can be determined from the relevant failure rate using equation 5.11. Many engineers have found it necessary and useful to employ a more compact approximation of this equation when performing a large number of iterative equations or when it is necessary to integrate the equation, as described in section 5.5.4. The simplified equation is derived by replacing the exponential function with the first two terms of its Taylor polynomial expansion. The Taylor polynomial expansion of the exponential function, to the nth degree, is shown in equation 5.14:

$$e^{x} = 1 + x + \frac{x^{2}}{2!} + \frac{x^{3}}{3!} + \ldots + \frac{x^{n}}{n!} + \frac{e^{z}}{(n+1)!}x^{n+1} \quad (5.14)$$

Rewriting equation 5.14 by using only the first two terms of the expansion and replacing x with $-\lambda t$ yields equation 5.15:

$$e^{-\lambda t} \approx 1 + (-\lambda t) \quad (5.15)$$

When we use the right side of equation 5.15 to replace the exponential function in equation 5.11, the result is equation 5.16, which is the approximation used for estimating probability based on mission time and test interval:

$$PFD_{Max} = 1 - e^{-\lambda t} \approx 1 - (1 + (-\lambda t)) \approx \lambda t \quad (5.16)$$

This approximation is effective when the values of the product of failure rate and mission time are small. Figure 5.3 shows the divergence of the values of equation 5.11 and the simplified equation 5.16 as the value of the product of mission time and failure rate grow. It is important to note that the value of the approximation is conservative, meaning that the probability of failure calculated by using the simplified equation will be greater than the probability calculated with the actual equation. As a guide for determining the accuracy of the approximation, remember that when a PFD of 0.3 is calculated using equation 5.16, the error between the approximation and the actual value is 15 percent (i.e., the actual value is a PFD of 0.26). This error drops to 5 percent when the PFD is 0.1. As the PFD falls below this number, the error continues to diminish. The calculations that are used for selecting SILs determine order-of-magnitude differences in the effectiveness of safety instrumented systems. For the purpose of those calculations, the simplification presented in equation 5.16 will be sufficiently accurate if the calculated PFD is less than 0.1.

5.6.2 Mutually Exclusive Probability Addition

We can greatly simplify probability addition by using the mutually exclusive form of the equation in most cases. The mutually exclusive and non-mutually exclusive forms of the probability addition equation are shown in equations 5.5 and 5.7, respectively. An inspection of the equations reveals that equation 5.5 requires only a simple addition, while equation 5.7 requires a more lengthy series of multiplications. Analysts have found that using the mutually exclusive form makes it possible to perform calculations much faster.

By examining equations 5.5 and 5.7 we can see that we gain a conservative result by using the mutually exclusive form in cases where the events are not mutually exclusive. The difference between mutually

Figure 5.3 Comparison of Probability Approximation to Actual

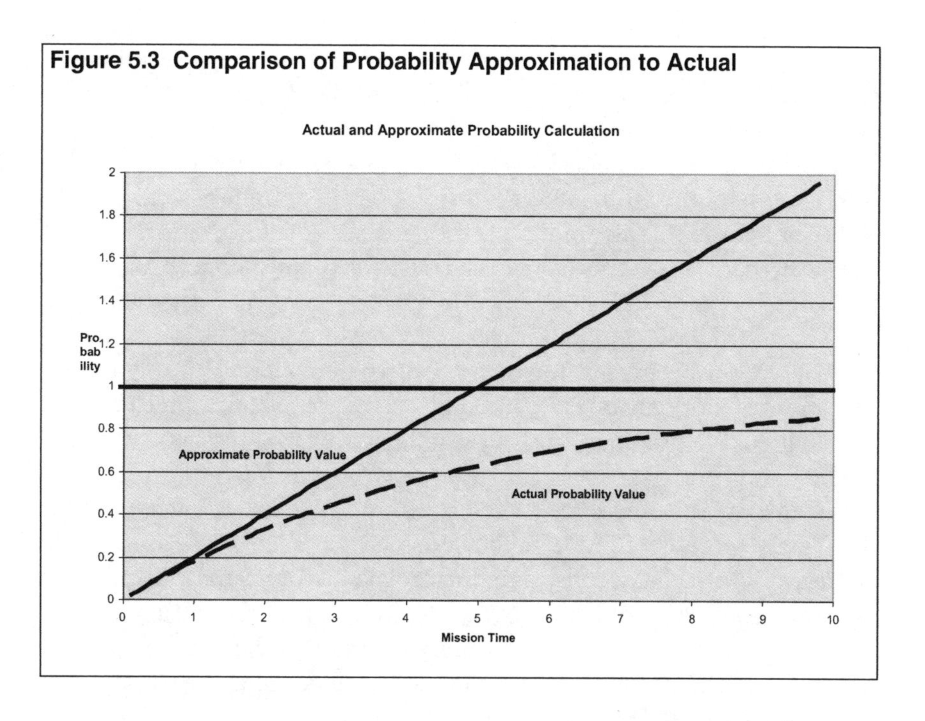

exclusive and non-mutually exclusive methods becomes very small as the event probabilities decrease. For instance, consider non-mutually exclusive events logically related by an OR whose individual probabilities are P=0.15. Using the mutually exclusive equation (5.5) gives the probability of the combination as P=0.2775. Using the non-mutually exclusive form (equation 5.7) yields a probability of P=0.3, which results in a value 8 percent too conservative. Similarly, if the individual event probabilities were P=0.05, then the error in using the mutually exclusive form of the probability addition equation will drop to 3 percent. As with the approximation presented in section 5.6.1, the simpler mutually exclusive form of the probability addition equation for non-mutually exclusive event sets is accurate enough for the purposes of selecting SILs if the calculated probability is less than 0.1.

5.7 Summary

This chapter presented the rules of probability and the ways they can be used to calculate the probabilities of individual events as well as combinations of events. We also presented an overview of the relationship between failure rates and probability of failure on demand (PFD). Probabilities are assigned to individual events by either examining the physical characteristics of a situation, such as the number of faces on a die, or by examining data, such as that based on operating history or equipment testing.

Probability math is dependent on the type of event under consideration. Independent events are not dependent on the outcome of other events. In reliability engineering, a single component failure is often assumed to be an independent event from all other component failures. But, because of the strong effect of common cause on the resulting probability of failure, this assumption must be approached with great care. Event sets can also be mutually exclusive and complementary. Mutually exclusive events occur when only one outcome in a set of possibilities can occur at any one time. Complementary events are a special case of mutually exclusive events in which only two outcomes are possible. In this instance, not only does the occurrence of one outcome preclude the possibility of the other, but their probabilities always sum to one because one and only one of the two outcomes must be true.

To calculate the outcome of a combination of events we need to know the type of the events and how they are logically related. When events are logically AND'ed, we calculate the probability of outcome by probability multiplication. When events are logically OR'ed we calculate the probability of outcome using probability addition. Two equations can be used for probability addition. The equation used depends on whether the events are mutually exclusive.

Fault tree analysis is used to describe and quantify complex relationships between different events. Fault tree analysis is a top-down approach in which basic events are related to the top event, whose probability is the object of study, through a series of logical gates. OR gate outputs are evaluated using probability addition, while AND gate outputs are evaluated using probability multiplication.

Quantification of equipment failure rates depends on historical information about how these events occurred in the past. The best source of this failure rate information is failure records and maintenance data on the specific equipment in the process plant under study. If plant data is unavailable, you can obtain information from several publicly available industry- and equipment-specific databases. You must often use careful engineering judgment when applying these databases and consider as many of the details of the individual case as possible. You must convert the failure rates presented in databases and calculated from plant equipment failures into probability of failure in order to use them in SIL selection calculations. The probability of failure is a function of failure rate, mode of failure, and the mission time between proof tests.

A number of simplifications and approximations are used to speed the risk analysis process and to derive expressions when the full forms of the equations are too cumbersome. When using these approximations, you should be careful to consider the potential amount of error that the simplification introduces as well as the direction of the error (i.e., does the approximation provide a conservative result?).

5.8 Exercises

5.1 Which probability action (multiplication or addition) is used to calculate the probability of a logical AND of two independent events?

5.2 What are the two methods for assigning probability for an event? Which method is used to determine the probability of drawing the ace of spades from a deck of cards?

5.3 If the failure and success of a relief valve are complementary events, and the probability of success is 99 percent, what is the probability of failure?

5.4 Consider an example of a potential process release. In this situation, a chemical is being pumped into a tank and could spill out of the tank's overflow discharge piping. For this situation to occur, (a) there would have to be more material in the delivery truck than space available in the tank—the probability of this occurring is 0.25, AND (b) the operator would have to forget to stop the pump when the tank is full—the probability of this occurring is 0.1. What is the probability that the tank is overfilled?

5.5 A pair of fair dice is rolled. What is the probability of getting a total roll of 4?

5.6 Draw and quantify a fault tree for the following situation. The flow of water to a process will be lost if the valve controlling the flow closes. The valve fails into the closed position when power is lost to its electronic actuator. There is also a possibility that the valve will fail into the closed position because the valve stem broke. The power to the valve's electronic actuator is supplied by two separate power sources, both of which would have to fail for the electrical supply to be lost to the valve. Over the course of one year, the probability of failure of power supply A is 0.2 and that of power supply B is 0.05. The probability of valve steam breakage is 0.01.

5.7 A plant has 200 identical solenoid valves, each of which is subjected to an annual function test. Over the course of 15 years, 75 dangerous failures have occurred. What is the failure rate and probability of failure on demand (maximum) for the solenoid valves? (Assume there are enough failures for a simple failure rate calculation to be valid.)

5.8 What is the mean time to failure (MTTF) of a component whose probability of failure on demand (average) is 0.05 with a test interval of three years?

5.9 Explain why failure rate data developed from a plant's internal maintenance databases will be more accurate than failure rates developed from sources such as equipment vendor returns and generic industry databases.

5.9 References

1. Edwards, C. H., and D. E. Penney. *Calculus and Analytical Geometry,* 2d ed. Englewood Cliffs, NJ: Prentice-Hall, 1986.

2. Goble, William M. *Control Systems Safety Evaluation and Reliability.* Research Triangle Park, NC: ISA, 1998.

3. Lees, F. P. *Loss Prevention for the Process Industries.* London: Butterworth and Heinemann, 1992.

4. Smith, David J. *Reliability, Maintainability, and Risk.* London: Butterworth and Heinemann, 1993.

CHAPTER 6

Consequence Analysis Overview

In this chapter we discuss the methods available for estimating consequences, including qualitative estimates, statistical analysis, and modeling the chemical and physical phenomena surrounding a release event. The release of flammable or toxic chemicals can have a variety of outcomes. The chemical release event tree is one tool that helps analysts determine the full range of potential consequences and plan any necessary emergency responses. This chapter focuses on the ways an event tree can be used to establish which event or events might occur based on the conditions that lead up to and surround a release.

This chapter will also provide an overview of qualitative, semi-quantitative, statistical, and quantitative methods for analyzing consequences. We will review a range of tools available to help the user with the most accurate method of consequence analysis, numerical chemical release phenomena modeling. Many consequence analysis tools are available, varying widely in their capabilities and ease of use. We discuss the limitations and strengths of these tools to help users select an appropriate tool for their own analysis.

The first step in determining the consequence of an unwanted event is defining a hazard zone, or area where receptors, such as people, equipment, environment, and businesses, will be harmed. The second step is determining the consequence based on that hazard zone. This step requires a knowledge of the vulnerability of people and property to the effects of exposure and also of the parameters for determining how likely they are to be impacted. These include capital density, personnel density, and the location of normally occupied buildings.

In general, the body of knowledge required to perform quantitative consequence analysis is very extensive and many of its particulars are beyond the scope of this text. Readers interested in performing detailed quantitative consequence analyses are encouraged to use third-party software tools that facilitate the process. More information on quantitative consequence analysis can be found in the textbook *Guidelines for Consequence Analysis of Chemical Releases* available from the Center for Chemical Process Safety (CCPS) of the American Institute of Chemical Engineers (AIChE). In this chapter we present an overview of the techniques of quantitative consequence analysis, introduce the important parameters and variables that impact consequences, and offer guidance on the tools and techniques useful in the estimation process.

6.1 Introduction to Consequence Analysis

Consequence analysis is the act of estimating the damage that results from a process accident. In the language of risk assessment as published in the CCPS Quantitative Risk Analysis Guidance, consequence is a measure of the expected outcome of an event. It is measured or expressed as *effect distances* or *effect zones*. Consequence analysis of accidents in the process industries typically involves analyzing the release of hazardous chemicals. This is normally done by using mathematical models and computer software to specifically address the chemical and physical phenomena of the release.

Impact analysis addresses the potential effects of the hazardous consequence on the surrounding people and property. Although traditional risk management focuses on the impacts to people, which include potential fatalities and other injuries, a good risk management program should take a broader view. Good impact analysis also takes into account losses from business interruption, property damage, damage to sensitive environments that can be adversely affected by chemical release, and third-party liability. The losses from a process incident can be tangible, such as property damage, or intangible, such as degraded corporate image. The common thread in all of these losses is that they all eventually impact the organization's financial bottom line. While the result of the chemical release modeling is an effect zone, the final consequence analysis is only completed after the impact analysis, which expresses the consequence in terms of loss of personnel or property.

The objective in selecting a safety integrity level (SIL) for a safety instrumented system (SIS) is to reduce the process risk to a tolerable level. The process risk is a function of both the likelihood that an event will occur and the associated consequences of that event. In some cases, process owners have made decisions based strictly on a hazard's likelihood, either ignoring the consequence altogether or performing only a rough qualitative estimate. Decisions based on this type of process often lead to poor results and ineffective resource allocation. Since the risk is a function of both consequence and likelihood, it is a waste to spend disproportionate effort determining likelihood because it is lost in the error generated by the poor consequence analysis technique. In the extreme case where the consequence is ignored altogether, the process owner unfortunately has no real idea of the amount of risk the process poses.

Quantitative models that evaluate the consequences of a hazardous chemical release are widely used in the loss prevention discipline. Organizations that incorporate quantitative consequence models into their

SIL determination method will consistently select the proper SIL. Chemical release models are particularly useful because the consequences of chemical releases are difficult to understand and often counterintuitive. Additionally, small changes in seemingly unimportant parameters can have large effects on the outcome. For instance, whether a toxic release occurs at 3 P.M. or 3 A.M. can change the consequence by an order of magnitude! (This is because solar surface heating dramatically affects both liquid volatility and atmospheric instability.) The amount of error resulting from the subjectiveness of qualitative estimates of consequence magnitude is often unacceptable for risk reduction engineering.

6.2 Methods for Performing Consequence Analysis

Consequence analysis is practiced in industry in a variety of ways. The method used depends on a number of factors, including:

- The organization's degree of experience with the hazard and with the process containing the hazard
- The level of sophistication of the engineering staff estimating the consequence
- The amount of data available on the impacts of past accidents

6.2.1 Qualitative Methods

Qualitative estimation is a procedure by which an expert or a team of experts estimates the consequence of a hazard by simply using judgment based on their personal and corporate experience with the process. For this type of analysis to be effective, the corporation and individual experts must have a large experience base. This approach usually results in rules of thumb for the type and amount of risk reduction required for familiar situations. Although this method has the strength of simplicity, its drawbacks include its reliance on historical accidents and a large, broad pool of experienced personnel. An organization that does not have a good "institutional memory" will often fail when trying to implement this strategy. Since this method is reactive to accidents that have already occurred, it is also ineffective for processes that are new or have not had a large number of accidents. Also, qualitative estimation tends to emphasize events subjectively and may not properly consider all of the different possible outcomes.

6.2.2 Semi-Quantitative Methods

Semi-quantitative risk indices, such as the Dow Fire and Explosion Index and the Dow Chemical Exposure Index, are good tools for developing a general feel for the amount of risk in a process. These indices use general process parameters to give the process a "score" that reflects a relative level of risk. Although this method is an improvement over purely qualitative methods, it has its own weaknesses, stemming from the fact that the score output is difficult to translate into an actual zone that would be impacted by an accident.

Sometimes these semi-quantitative methods provide a result that combines the estimate of the consequence zone and the impact analysis within that zone. Although this is typically better than a purely qualitative result, it can significantly hamper the later processes of designing an appropriate SIS and non-SIS protection structure.

6.2.3 Statistical Analysis of Accidents

Statistical analysis is an excellent tool for consequence analysis, as long as sufficient data is available, which is rarely the case for process accidents. In essence, the consequence of an event is the statistical average of the consequences of all prior comparable events. This method is very effective when the data set is large, such as for motor vehicle accidents. It is not as effective in situations such as chemical releases where sufficiently large and specific data sets are unavailable.

Consequences are estimated through statistical analysis that uses a variation of equation 5.1. Essentially, the average consequence is determined by summing the total consequences and dividing by the total number of accidents. In other words, we can assume that the consequence of an event is equal to the mean loss of all of the previous loss events. If the goal of an analysis is to determine the probable consequence, in terms of fatalities, for a car crash, the analyst should divide the total number of fatalities caused by car crash for a certain time interval by the total number of car crashes during that same interval.

The applicability of this method is narrow because extensive historical data must be amassed. It is not enough for the data set to cover a large time period; it must also contain a large number of accidents. In addition, for the consequences of the past accidents to predict future accidents, the circumstances under which the accidents occur must be nearly identical. This method is strong for estimating the consequences of events such as vehicle accidents. Most automobiles are roughly the same in design, and the speeds at which they travel and the number of passengers they carry are roughly the same. Therefore, an automobile insurance company can estimate with great accuracy the approximate consequence of a future automobile accident and then use this information to determine insurance premiums. Unfortunately, statistical analysis is usually not effective for determining the consequences of process plant accidents. For the same reasons, it is also ineffective for determining the likelihood of process plant accidents. The particular details and characteristics of each individual process plant are sufficiently different that general conclusions about consequences are usually not valid. Additionally, the number of process plant accidents is not large enough to provide a base for any valid statistical conclusions.

Example 6.1

Problem: In the United States in 1998, there were 6,842,000 motor vehicle accidents. As a result, there were 41,907 fatalities and 3,511,000 injuries. Given this information, compiled in the Bureau of Transportation Statistics Annual Report, what are the consequences, in terms of injuries and fatalities, of a motor vehicle crash?

Solution: The two terms used to describe an event's consequences in terms of injuries and fatalities are probable injuries (PI) and probable loss of life (PLL), respectively. PLL and PI are both calculated using equation 5.1 for assigning probability based on trial or historical data:

For fatalities, PLL = 41,907 / 6,842,000 = 6.12×10^{-3}

For injuries, PI = 3,511,000 / 6,842,000 = 0.513

Based on this information, if you are involved in a motor vehicle crash (that is reported to DOT) there is approximately a 50 percent chance of someone involved in the crash sustaining an injury and one chance in ~163 of someone being killed.

6.2.4 Quantitative Methods: Release Phenomena Modeling

When statistical analysis is not appropriate, release phenomena modeling can be used to predictively model the consequences of a hazard. Release phenomena modeling works by first analyzing the potential energy that a hazard contains in its pre-accident state. The method then estimates the effect of the release of that energy under the conditions that result from the loss of control of the process.

In general, a hazard stems from an uncontrolled release of potential energy. This potential energy, when contained by the process, provides useful work. When this potential energy is released to the atmosphere and allowed to contact unanticipated receptors, a range of harmful impacts may result. It is important to understand that the energy potential is not limited to any one type. It includes mechanical energy, thermal energy, chemical reactive potential, and electrical potential. In process plants, this release falls into three main categories: physical (such as a high-pressure vessel rupture), flammable, or toxic.

Modeling the physical and chemical phenomena surrounding a release is a very effective method, but it is also time consuming and requires a high level of expertise. In some situations, this method is the only viable option. Software tools are available for performing chemical release modeling to varying degrees of mathematical and numeric rigor. Some of the more common ones are listed in table 6.1 of section 6.6.

6.3 Consequence Definitions and Measures

Consequence is the end measurement of the effect of an unwanted accident. The consequence is always presented in terms of loss or quantity of harm sustained by a class of receptor. To understand consequence analysis, it is important to understand the terms that describe the event sequence that leads from a hazard to a consequence. The terms presented here are derived from the definitions used in *Guidelines for Chemical Process Quantitative Risk Analysis* (CPQRA). While the safety instrumented system standards, such as IEC 61511 and IEC 61508, present terms that overlap somewhat with the terms in CPQRA, the two sources of definitions should be used for separate purposes. The terms in the SIS standards are good for describing SIS design and analyzing the reliability and availability of safety instrumented systems. However, they are inadequate for analyzing chemical process risk.

The starting point is a hazard, which is "an accident waiting to happen." CPQRA defines a hazard in the following terms:

> ***Hazard:*** *A chemical or physical property that has the potential for causing damage to people, property, or the environment (e.g., a pressurized tank containing 500 tons of ammonia).*

The first step in the sequence between a hazard, which is contained and controlled by the process, and the unwanted accident, is the initiating event. CPQRA defines an initiating event as follows:

> ***Initiating Event:*** *The first event in an event sequence (e.g., the stress corrosion resulting in leak/rupture of the connecting pipeline to the ammonia tank).*

Although the CPQRA example is of an initiating event in which a significant outcome will occur, this is not always the case. Consider an initiating event where the fuel gas supply to a fired heater is temporarily contaminated with nitrogen, causing a loss of flame at the burner. If loss of flame is detected and the fuel gas flow to the burner is then stopped, no significant outcome will result from this initiating event.

An intermediate event can be an event in the series between the initiating event and the final outcome. As with the initiating event, the chain leading to an unwanted accident can also be broken with the intermediate event and no harmful outcome need result. In the example of a gas burner's loss of flame, without a fuel cut-off action, the gas could build up to significant explosive levels. This buildup would be an intermediate event. The event chain could stop at this intermediate event if the surfaces immediately surrounding the burner are hot enough to cause rapid re-ignition of the gas before enough can accumulate to cause a severe explosion. Obviously, one would not want to rely on this as a substitute for an SIL 4 SIS, but breaking the chain of propagation at an intermediate event will mitigate or prevent the harmful outcome. CPQRA defines an intermediate event in this way:

> ***Intermediate Event:*** *An event that propagates or mitigates the initiating event during an event sequence (e.g., improper operator action fails to stop the initial ammonia leak and causes propagation of the intermediate event to an incident; in this case the intermediate event outcome is a toxic release).*

An incident is the result of an initiating event that is not prevented from propagating. The incident is the most basic description of an unwanted accident, and provides the least information. The term *incident* is simply used to convey the fact that the process has lost containment of the chemical or other potential energy source. Thus, the potential for causing damage has been released, but its harmful result has not yet taken specific form. CPQRA defines incident as follows:

> ***Incident:*** *The loss of containment of material or energy (e.g., leak of 10 lb/s of ammonia from a connecting pipeline to the ammonia tank, producing a toxic vapor cloud); not all events propagate into incidents.*

When a loss-of-containment event, or incident, has already occurred, a number of different consequences are often possible. For instance, the release of acrylonitrile could result in a pool fire, flash fire, unconfined vapor cloud explosion, toxic effect, or a combination of these effects. The term used to describe the particular effect being analyzed is *incident outcome*. CPQRA defines incident outcome in the following terms:

> ***Incident Outcome:*** *The physical manifestation of the incident; for toxic materials, the incident outcome is a toxic release, while for flammable materials, the incident outcome could be a Boiling Liquid Expanding Vapor Cloud Explosion (BLEVE), flash fire, unconfined vapor cloud explosion, toxic release, etc. (e.g., for a 10 lb/s leak of ammonia, the incident outcome is a toxic release).*

Each incident can have a number of incident outcomes. Similarly, in a second level of classification, each incident outcome can have a number of incident outcome cases. For instance, a release can occur under a variety of weather conditions, and each of those weather conditions will result in a different magnitude of consequence. As such, a number of difference incident outcome cases are required to describe the same incident outcome. One incident outcome case could be used for each weather condition under study. So a single incident such as a chemical release can have several incident outcomes such as a "release / vapor cloud explosion," and each incident outcome can have several incident outcome cases such as a "release / vapor cloud explosion / north wind" case that causes the vapor cloud explosion to miss the main control room. CPQRA defines incident outcome case as follows:

> ***Incident Outcome Case:*** *The quantitative definition of a single result of an incident outcome through specification of sufficient parameters to allow distinction of this case from all others for the same incident outcomes. For example, a release of 10 lb/s of ammonia with a D atmospheric stability class and 1.4 mph wind speed gives a particular concentration profile, resulting, for example, in a 3000 ppm concentration at a distance of 2000 feet.*

For each incident outcome case being considered, the risk analysis will determine the resulting effect zone. The effect zone is an area where a measurable magnitude of the incident outcome is above a certain threshold. For instance, the magnitude of a fire is proportional to the heat radiation produced by that fire, usually measured in kW/m^2. The actual value of the criterion that is used, which is called the *endpoint*, is typically selected to represent a certain level of vulnerability that a receptor has to that incident outcome. For example, the U.S. EPA selected a thermal radiation endpoint of 5 kW/m^2 to represent an injury endpoint when formulating its Risk Management Program regulation. CPQRA defines an effect zone as follows:

> ***Effect Zone:*** *For an incident that produces an incident outcome of toxic release, the area over which the airborne concentration exceeds some level of concern. The area of the effect zone will be different for each incident outcome case [e.g., given an IDLH for ammonia of 500 ppm (v), an effect zone of 4.6 square miles is estimated for a 10 lb/s leak]. For a flammable vapor release, the area over which a particular incident outcome case produces an effect based on a specified overpressure criterion (e.g., an effect zone from an unconfined vapor cloud explosion of 28,000 kg of hexane assuming 1% yield is 0.18 km^2 if an overpressure criterion of 3 psig is established). For a loss-of-containment incident producing thermal radiation effects, the area over which a particular incident outcome case produces an effect based on a specified damage criterion [e.g., a circular effect zone surrounding a pool fire resulting from a flammable liquid spill, whose boundary is defined by the radial distance at which the radiative flux from the pool fire has decreased to 5 kW/m^2 (approximately 1600 $BTU/hr\text{-}ft^2$)].*

Once we have established the effect zone, we can determine the consequence by estimating the number of receptors that will receive the incident outcome case's effects. The consequence is simply a measurement of the impact, in terms of loss, of the incident outcome case. CPQRA defines consequence in this way:

> ***Consequence:*** *A measure of the expected effects of an incident outcome case (e.g., an ammonia cloud from a 10 lb/s leak under stability class D weather conditions, and a 1.4-mph wind traveling in a northerly direction will injure 50 people).*

While this definition stresses the fact that consequence is determined for an individual outcome case, the problem of SIL selection usually requires that planners consider an overall or average outcome for all potential incident outcome cases stemming from a single incident. Event tree analysis can be used to calculate an average consequence for an incident. This procedure will be discussed in chapter 8.

Both effect zones and consequences will vary depending on the type of receptor. For instance, the effect zone surrounding a fire will be larger for injuries than for fatalities. To state a consequence properly and clearly, we must include both its size and the type of receptor, or damage. The receptors typically considered when performing risk analyses include injuries to people, loss of life, equipment damage, environmental damage, and business interruption.

Any of these types of damage, or receptors, can be used to represent consequences depending on the type of decision that is involved. For instance, decisions about personnel safety usually require that the consequence be stated in terms of both injuries and fatalities. However, cost-benefit analysis of equipment protection systems require that consequence be stated in terms of property damage and financial cost.

Optimally, all of the types of losses should be considered simultaneously. This requires that conversion factors be used to determine the relative utility of each of the loss types so the proper trade-off can be made between them. Using this procedure, we can derive a single uniform basis of loss that will facilitate the decision-making process by making the decision one-dimensional.

6.4 Quantitative Analysis of Chemical Releases

Before conducting a consequence analysis, it is necessary to identify incident outcomes that could follow the release of any hazardous chemicals involved in the process. Figures 6.1, 6.2, and 6.3 show simplified event trees for the release of hazardous chemicals: for a general initial loss of containment, for a gas release, and for a liquid/liquefied gas release, respectively.

When control of a chemical process is lost, either containment of the chemicals in the process will be lost or the process will correct itself and come to a safe state with no release of chemical and no consequence. If a

Figure 6.1 Event Tree for Simplified Loss of Chemical Containment

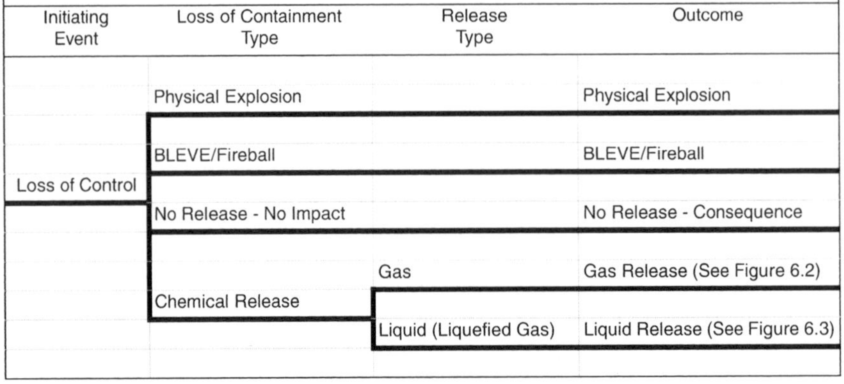

Initiating Event	Loss of Containment Type	Release Type	Outcome

Loss of Control
Physical Explosion — Physical Explosion
BLEVE/Fireball — BLEVE/Fireball
No Release - No Impact — No Release - Consequence
Chemical Release — Gas — Gas Release (See Figure 6.2)
Chemical Release — Liquid (Liquefied Gas) — Liquid Release (See Figure 6.3)

Figure 6.2 Event Tree for Gas Release

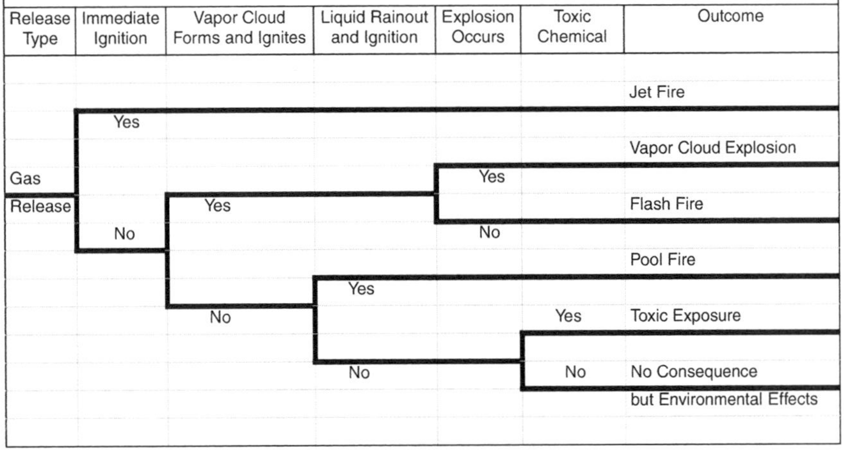

Release Type	Immediate Ignition	Vapor Cloud Forms and Ignites	Liquid Rainout and Ignition	Explosion Occurs	Toxic Chemical	Outcome

Gas Release
Yes — Jet Fire
No — Yes — Yes — Vapor Cloud Explosion
No — Yes — No — Flash Fire
No — No — Yes — Pool Fire
No — No — No — Yes — Toxic Exposure
No — No — No — No — No Consequence but Environmental Effects

chemical release does occur, the consequence may result directly from the release event, as is the case with a BLEVE/fireball or physical explosion (see section 6.4.1). Alternatively, the chemicals will be released to the atmosphere and potentially cause damage later depending on their properties.

The two consequences that are coincident with the initial release event are physical explosions and boiling liquid expanding vapor explosions (BLEVEs), with their corresponding fireballs. If a pressure vessel is stressed beyond its design pressure, it can undergo a catastrophic failure, creating a physical explosion. These are often euphemistically referred to as "energy releases" when speaking with the media. If the material released as the result of a physical explosion is flammable, a fireball may also accompany the physical explosion. If the accident involves a flammable liquid spill, followed by ignition, and a complete flame engulfment of a flammable tank, a BLEVE may occur. If the loss of containment event does not cause an immediate coincident effect, the chemical contained in the process will be released into the atmosphere. That chemical may be involved in a variety of effects depending on the release conditions and properties of the specific compounds released. The thermodynamic state of the released material, whether it is liquid, gas, or liquefied gas, will have a large impact on what incident outcomes are possible.

If the released material is a gas or a high-pressure liquid that instantly flashes into a gas upon release, a jet fire will result if the gas is immediately ignited. In the absence of immediate ignition, a large flammable vapor cloud may form. Delayed ignition of the cloud may lead to a vapor cloud explosion (VCE) with its attendant blast overpressure and shock wave. Depending on the characteristics of the released material and the surroundings, a vapor cloud may not result in an explosion even after ignition. In this case, the cloud will burn in a slightly slower laminar fashion, causing a flash fire, which has significant thermal effects, but does not cause a blast wave. The difference between the two depends on the complicated phenomenon of flame propagation velocity, which requires complex modeling to predict with any accuracy. Even if there is no ignition of any kind, the unignited cloud of potentially toxic gas will spread and disperse, which could adversely affect workers and nearby residents. Unignited gas releases, and in some cases the combustion products of ignited releases, can also have a detrimental effect on the surrounding environment.

Figure 6.3 Event Tree for Liquid Release

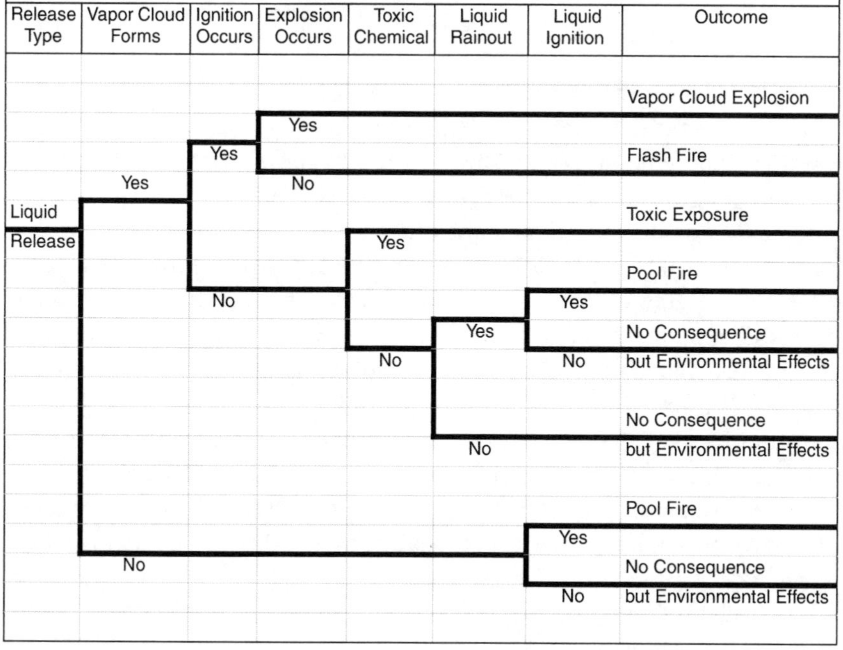

Release Type	Vapor Cloud Forms	Ignition Occurs	Explosion Occurs	Toxic Chemical	Liquid Rainout	Liquid Ignition	Outcome

Liquid Release
Yes
Yes
Yes
Vapor Cloud Explosion
No
Flash Fire
No
Yes
Toxic Exposure
No
Yes
Yes
Pool Fire
No
No Consequence but Environmental Effects
No
No Consequence but Environmental Effects
No
Yes
Pool Fire
No
No Consequence but Environmental Effects

The incident outcomes that might occur as the result of a liquid or a liquefied gas discharge are largely a function of the behavior of the liquid upon release. One of three scenarios is possible after the release of liquid from a process: (a) immediate and complete flash of liquid to vapor, (b) rapid flashing of liquid to vapor with substantial formation of a liquid pool, or (c) slow to negligible flashing to vapor with significant liquid pooling. If the liquid immediately and completely flashes to vapor upon release, the event tree shown in figure 6.2 will unfold. Otherwise, the event tree in figure 6.3 better represents the possible outcomes of the release.

Figure 6.3 shows that the outcomes from a liquid release involving vapor cloud formation are largely the same as they would be if a vapor cloud were formed directly from a vapor release. This cloud formation can result from either rapid vaporization or slow evaporation of a liquid pool. In the case where a pool of liquid is formed and ignited, a pool fire will result. If the pool is not ignited, evaporation of the contents of the pool may lead to a harmful exposure hazard downwind if the material is toxic. The material may also cause environmental damage such as groundwater contamination even if it is not ignited.

In both the vapor and liquid release cases, a potential exists that released material will be carried away from the release source as an aerosol or as a gas cloud, which will then cool, collect, and "rain out" of the atmosphere to collect in a pool. The hazards associated with a pool formed from rain out are essentially the same as the hazards from the direct spill of a liquid, except that they are some distance from the leak source. As a result, secondary containment would likely not help mitigate their consequences.

6.4.1 Flammability Effects

The incident outcomes of pool fires, jet fires, flash fires, and fireballs/BLEVEs all pose flammability hazards, that is, hazards in which harm can come to people and property because of the thermal energy released when a flammable material combusts. The thermal energy released from a fire is primarily radiated from the portions of the flame that are in a direct line of sight of the receptor, not obscured by smoke or other potentially shielding pieces of equipment. It is important to note that not all fires have visible flames. Hydrogen fires in the daytime, which burn with a very intense radiative heat, are one case in point. The radiation heat-transfer mechanism dominates the flammability hazard's ability to cause damage. Although conduction and convection effects are usually negligible, they can play a potential role when the hot gas combustion products are blown toward elevated structures by any wind that may be present during the incident.

The ability of a fire to injure people and damage property is a function of the thermal radiation that is absorbed by the receptor. The dose absorbed by a receptor will vary with location, orientation toward the flame surface, amount of smoke produced, humidity, and other atmospheric conditions. Sheltering behind equipment, in buildings, and behind purpose-built thermal radiation shields can reduce the magnitude of a thermal radiation dose. The consequence of a fire is typically described in terms of the distance (endpoint) to a specific level of thermal radiation flux, usually measured in kW/m^2. For instance, World Bank figures indicate that a direct skin exposure to 5 kW/m^2 of thermal radiation for 40 seconds will result in serious third-degree burns.

6.4.1.1 Pool Fire

Spilled flammable liquids will create a pool fire if ignited. The magnitude of the effect zone created by a pool fire will depend on the size of the flame it generates, which in turn depends on the size of the spill surface and the properties of the spilled fluid. The flame's footprint is determined by the containment of the liquid spill, which is often controlled by any dikes or curbs present. If a spill is unconfined, the liquid will spread over an area determined by the fluid's viscosity and the characteristics of the surface on which the material is spilled, such as its porosity.

The height of the flame is determined by the characteristics of the burning fluid. A liquid's vapor pressure and heat of vaporization will determine the rate at which the liquid will volatilize and contribute to the oxidation reaction. Higher vapor pressures and low heats of vaporization cause more vaporization to occur, and thus faster reaction and more energy release. The combustion energy of the liquid determines the amount of energy that is released per unit of liquid. Other properties such as flame speed and adiabatic flame temperature are also important in determining the thermal effects of a pool fire.

How completely a material is combusted will determine the amount of smoke a fire generates. The amount of smoke generated is important because a fire's energy is only radiated from the visible portions of a flame. If smoke is obscuring a significant portion of the flame, the transmitted energy will be greatly decreased. For instance, if a diesel fuel pool fire and a liquid natural gas (LNG) pool fire have the same dimensions, the LNG fire will have a much larger effect zone. The reason the LNG fire will have a larger effect zone is that diesel fires are very smoky, and the smoke obstructs the radiative transmission of energy. Atmospheric factors such as wind speed can also affect a pool fire's height by causing the flame to tilt.

Another important effect of incomplete combustion is the toxic nature of many of the partially burned compounds that can be formed. Although complete combustion releases more heat, the toxic effects of the primary CO_2 and H_2O combustion products is minor. However, the soot and other various toxic partial oxidation products from incomplete combustion can widely disperse in a pool fire, leading to many other potentially serious impacts.

A more detailed discussion about modeling the effects of pool fires, including analysis of the ways thermal effects impact humans and structures, can be found in section 2, chapter 4, of the *NFPA/SFPE Handbook of Fire Protection Engineering* by Mudan and Croce (National Fire Protection Association and Society of Fire Protection Engineers).

6.4.1.2 Jet Fire

A jet fire results when high-pressure flammable material is ignited as it is being released from containment. The kinetic energy of the physical release thus helps both to mix the material with the oxygen in the air and to spread the resulting flame. The size of the flame is determined mainly by the physical conditions surrounding the release. As a high-pressure material is released from a hole, it will exit with a velocity that is mainly a function of system pressure and the hole's size. The greater the distance away from the hole, the more oxygen is present in the mixture as air is entrained in the jet. As the upper flammability limit threshold is crossed, the fuel and air react, releasing the energy of combustion. As the combustion continues, entrained air, unburned fuel, and combustion products continue to move in the direction of the release because of the momentum generated by the release.

The effect zone of a jet fire (also known as a torch fire) is determined, like that of a pool fire, by a combination of the physical characteristics of the release and the chemical properties of the burning material. The effect zone of a jet fire is proportional to the size of the flame that is generated.

We must also account for the chemical properties when determining the effects of a jet fire. Properties such as heat of combustion also play a factor here, but to a lesser extent than the release rate. Mudan and Croce present a detailed discussion of jet fire analysis in section 2, chapter 4, of the *NFPA/SFPE Handbook of Fire Protection Engineering*.

6.4.1.3 Flash Fire

A flash fire is the result of the ignition of a cloud of flammable vapor when the flame velocity is too slow to produce an explosive shock wave. When a gas phase mixture of fuel and air is ignited, a flame front travels from the point of ignition in all directions where the fuel-air concentration is within flammable limits. The velocity of the flame front will determine the type of damage this event causes.

If the flame front burns in a laminar fashion, with the flame front traveling at less-than-sonic velocity, a flash fire will be the result. (The conditions under which the same scenario would result in a vapor cloud explosion are described in section 6.4.2.) If the flame front does not reach sonic velocity, then an overpressure condition does not develop. The consequences in the flash-fire scenario are mainly due to the heat of combustion being absorbed by receptors in the effect zone. For a flash fire, the effect zone is limited to the "flame envelope," or the area where the fuel-air mixture is within the flammable limits. We can determine the size of the flame envelope using dispersion modeling (see section 6.4.3, "Toxic Hazards: Dispersion Modeling," for more information on vapor dispersion modeling).

A flash fire will not develop the overpressure shock wave that is caused by a vapor cloud explosion. Thus, there will be no equipment damage caused by shock waves or other damage caused by projectiles. In addition, the duration of a flash fire is very short compared to a pool fire or jet fire. As a result of the short duration, the amount of equipment damage in the flash fire effect zone will be small. Although the equipment damage caused by a flash fire will be limited when compared to a vapor cloud explosion, human impacts can be severe. It is quite likely that any persons in the flammable envelope when a flash fire occurs will be fatally injured.

6.4.1.4 Fireball/ BLEVE

A fireball is the result of a sudden and widespread release of a flammable gas or volatile liquid that is stored under pressure coupled with an immediate ignition. It is distinguished from a jet fire by the shorter duration of the event and the difference in the geometry and shape of the flame. When a pressure vessel containing a flammable gas or volatile liquid ruptures, the first result is the quick dispersion of the flammable material as the highly pressurized material rapidly expands to atmospheric pressure. During this expansion, the release will entrain large quantities of air. If the material in the vessel is a volatile liquid, this process will also cause an aerosol to form with the dispersion of liquid droplets away from the release as a result of the vapor expansion.

Shortly after the initial release, the expanding vapor cloud will entrain enough air to reach the upper flammability limit. If at this point a source of ignition is present, the vapor cloud will rapidly combust. Ignition sources are common after catastrophic pressure vessel ruptures because of flying metal fragments and the heat generated by the rupture process. As the ignited cloud combusts, it continues to expand further. The combination of an expanding flame front; clean, relatively smoke-free combustion; and rapid reaction creates a fire that travels a substantial distance away from the release source and produces intense heat. As the cloud nears the latter stages of its burn, the density of the fireball drops because of the high temperature of the combustion products. When this occurs, the cloud becomes buoyant and lifts off the ground. This is the reason for the "mushroom cloud" that always accompanies a fireball.

A substantial amount of research has been done into fireballs including both theoretical work and empirical correlation of industrial accidents. In section 16.15 of his *Loss Prevention for the Process Industries,* Lees presents a comprehensive analysis of the series of equations and relationships that are required to model a fireball.

A boiling liquid expanding vapor explosion (BLEVE) is a specific type of fireball, but the two are not synonymous. While all BLEVEs result in fireballs, not all fireballs are the result of BLEVEs. BLEVEs occur when vessels containing a pressurized liquid come in direct contact with external flame. This contact can result from the vessel being engulfed in flame or from a jet fire impinging onto the vessel surface. As the liquid inside the vessel absorbs the heat of the external fire, the liquid begins to boil, increasing the pressure inside the vessel to the set pressure of the relief valve(s). The heat of the external fire will also be concentrated in the portions of the vessel where the interior wall is not "wet" with the process liquid. Since the process liquid is not present to carry heat away from the vessel wall, the temperature in this region (usually near the interface of the boiling liquid) will rise dramatically. This will cause the vessel wall to overheat and become weak. A short time after the vessel wall begins to overheat, the vessel will lose its structural integrity and a rupture will occur. After the vessel ruptures, a fireball, as described previously, will usually result, ignited by the external fire.

6.4.2 Explosion Effects

The consequences of an explosion hazard are related to the effects caused by the explosion's blast wave. On a fundamental level, the blast wave is simply a thin shell of compressed gas that travels away from the source of the explosion as a three-dimensional wave. The magnitude of the blast wave is typically defined by its peak overpressure, or the difference in pressure between the highest pressure point in the "shell" and ambient atmospheric pressure. A blast wave also has other parameters that describe its effect, such as duration and impulse, but the simple use of peak overpressure is the most common method for describing and classifying explosion effects.

The correlation of explosion parameters, such as a peak overpressure, to the damage sustained by persons, equipment, and structures has been the subject of a great amount of detailed study. Reviews of accidental explosions and explosion studies have shown that a 5.0-psi overpressure can cause substantial damage to most typical process equipment, and as little as a 0.5-psi overpressure can cause glass breakage. Projectiles and collapsing buildings are the main contributors to an explosion's impact on people. The methods for estimating the vulnerability of humans and structures to explosion effects can be found in American Petroleum Institute's Recommended Practice 752- *Management of Hazards Associated with Location of Process Plant Buildings, The Effects of Nuclear Weapons,* and the

CCPS's *Guidelines for Evaluating Process Plant Buildings for External Explosion and Fire.*

6.4.2.1 Vapor Cloud Explosions

As we discussed in section 6.4.1.3, the ignition of a flammable fuel-air mixture cloud will either cause a flash fire or a vapor cloud explosion. While a flash fire results from a laminar flame front that is slower than the speed of sound, a vapor cloud explosion results from a flame front that is turbulent and exceeds sonic velocity. The explosion potential of a flammable release depends on the properties of the released material, the energy of the ignition source, and the confinement and obstacle density in the area of the release. Flame turbulence is typically formed by the interaction of the flame front and obstacles such as process structures or equipment. As the location of a vapor cloud explosion becomes more congested and confined, the likelihood of an explosion will increase.

In general, four primary conditions are required for a vapor cloud explosion. Preventing these conditions will usually prevent the explosion. First, the material must be released in the appropriate temperature and pressure range. Second, the ignition must be delayed enough to allow the fuel and oxidant materials to mix. Third, a sufficient fraction of the cloud must be in the flammable range, with more homogeneous mixtures causing stronger explosions. Finally, there must be a mechanism for generating turbulence, which could include the release itself or external turbulence induced by objects in the area. These four elements are important both for estimating the consequence of a vapor cloud explosion and for designing a method of protection against one.

The blast effects produced by vapor cloud explosions vary greatly and depend primarily on the resulting flame speed. Highly reactive materials such as acetylene and ethylene oxide are much more likely to lead to a vapor cloud explosion than low-reactivity materials such as propane because they can produce higher flame speeds.

Many models have been proposed and used for analyzing the effects of explosions. They range from the simplistic single-point TNT equivalency model to three-dimensional computational fluid dynamics that consider the attenuation and reflection of the blast wave due to obstacles in the blast path. The explosion models most commonly used for the rough explosion magnitude estimates required for selecting SILs include TNT equivalency, TNO multi-energy, and Baker-Strehlow. A complete overview of the methods for explosion modeling can be found in *Loss Prevention for the Process Industries.*

6.4.2.2 Physical Explosions

Explosions can be caused either by the ignition of flammable materials, as discussed previously, or by the sudden catastrophic rupture of a high-pressure vessel. The blast wave created by a high-pressure vessel rupture is often called a *physical explosion*. Although the causes of these explosions are different, the effects are essentially the same. In a physical explosion, the blast wave is caused when the potential energy that is stored as high pressure in the vessel is transferred to kinetic energy when the material stored in the vessel is released. A fireball may also occur if the released material is flammable and is immediately ignited.

A discussion of how to model the effects of a physical explosion can be found in CCPS's *Guidelines for Chemical Process Quantitative Risk Analysis*. The most common model for physical explosions is the same as the TNT equivalency model, except that it uses an alternate method to determine the energy that contributes to the blast wave. When performing a TNT equivalency analysis for a flammable material, the heat of combustion of the material is used to determine the amount of energy released. When performing the same analysis for a physical explosion case, the amount of work required to compress the gas from ambient conditions to the conditions under which the release occurs is assumed to be the energy contributing to the blast.

6.4.3 Toxic Hazards: Dispersion Modeling

Although the release of a toxic chemical may produce little, if any, property damage, it may result in a significant impact on the workforce and any surrounding off-site population. The effect of a toxic release will be caused by the biological reactivity of the toxic chemical, and not by any primarily energetic reaction that occurs. As such, we can identify the effects of a toxic chemical release by first determining what concentrations of the material will be present in areas downwind of the release and then what biologically toxic effects these concentrations have. The analysis of the concentrations of materials downwind of releases is called *dispersion modeling*.

Toxic effect zones are determined by the following parameters:

- Release quantity
- Duration of release
- Source geometry
- Elevation/orientation of the release
- Initial density of the release
- Prevailing atmospheric conditions
- Surrounding terrain
- Limiting concentration (endpoint)

We briefly discuss the most critical parameters in the following paragraphs, with special emphasis on their influence on the process of estimating the distance of downwind dispersion effects. A discussion of modeling the dispersion of a chemical release can be found in *Guidelines for Chemical Process Quantitative Risk Analysis*.

The release quantity refers to the quantity of a hazardous chemical that is released when an accident occurs. The release quantity is the single most important factor in determining dispersion effect distances. If the duration of the release is long, you may consider release rate instead of release quantity. In general, larger quantities lead to larger dispersion distances. However, the dispersion distance does not increase linearly with quantity or release rate. For gaseous and liquefied gas releases, the vapor release rate will be the same as the discharge rate. However, for liquids, the vapor release rate is governed by the evaporation rate of the liquid and will always be less than the total liquid release rate.

The duration-of-release parameter depends on the situation that causes the release as well as the physical characteristics of the release. Most dispersion models use one of the following two extreme cases: the release is assumed to either occur continuously, in which case the material is released at a constant rate for a long time, or instantaneously, in which case the entire quantity of material is released at once. Under the instantaneous release assumption, the duration of release should be very short (e.g., pressurized storage tank rupture), and the total quantity of the chemical released during the accident contributes to the dispersion hazard. Under the continuous release assumption, the release rate is the most important parameter because the downwind concentration profile of the released material will come to a steady state. The continuous addition of more material will only maintain the concentration profile at constant levels, but will not extend it further downwind.

The atmospheric conditions that impact the effect zone of a toxic release include atmospheric stability and wind speed. A lower wind speed usually leads to slower dilution and, hence, larger hazard areas. Lower wind speed also means that the vapor cloud will travel more slowly and take longer to establish a steady-state concentration profile. Atmospheric stability refers to the vertical mixing of the air to disperse a released chemical. These are classified as Pasquill Stability Classes, which range from A (highly unstable) to F (highly stable). The late afternoon hours are typically categorized as A or B, whereas the calm hours of the night or early morning are usually in the E or F categories. The stability determines the speed of dispersion, and therefore F stability usually leads to very large dispersion distances because very little vertical mixing is occurring. With everything equal, the difference in the dispersion distance for F stability and A stability can easily be an order of magnitude. The prevailing wind direction at the time of any release will determine in what direction a vapor cloud will move as well as the specific population and property that may be impacted.

The limiting concentration, or endpoint, is the cutoff point for the parameter of interest, usually the point where effects such as injury or death are expected to end. For instance, when modeling a release of hydrogen sulfide, an analyst might wish to determine the size of the effect zone in which the value Immediately Dangerous to Life and Health (IDLH) is exceeded. In this case, the limiting concentration would be 100 ppm. As one would expect, limiting concentration affects the dispersion distance inversely, with lower limiting concentrations leading to larger dispersion distances. As with source release rate and dispersion distances, the effect is not linear.

The benchmarks used to determine the effect of toxic chemicals include Emergency Response Planning Guidelines, or ERPGs, which were established by the American Industrial Hygiene Association (AIHA). Another such set of benchmarks is the Immediately Dangerous to Life and Health (IDLH) levels suggested by the U.S. National Institute of Occupational Safety and Health (NIOSH). It is important to make a distinction between concentrations at which there will be some observable effect and concentrations at which one can expect serious ill effects and potential fatalities. Typically, the concentrations at which one can expect fatalities are significantly higher (nearly 100 times in some chemicals) than the suggested ERPGs or IDLHs.

Information on the toxic effects of various compounds can be found in many different places. One basic starting point is the Material Safety Data Sheet (MSDS) for a substance. This information is required by law in the United States, and it can be readily found in databases accessible through the World Wide Web at numerous publicly accessible sites, including http://siri.uvm.edu/ maintained by the University of Vermont and the Vermont Safety Information Resources Inc. Specific IDLH database information as well as other toxicity data can be found through the U.S. National Institute for Occupational Safety and Health Web site at http://www.cdc.gov/niosh/database.html.

6.5 Effect Zone and Consequence

The consequence of an incident will depend on the size of the incident's effect zone and the zone's occupancy. Once both of these factors are known, you can integrate them into such consequence measures as Probable Loss of Life (PLL), Probable Injuries (PI), and Expected Value of loss (EV).

When determining the occupancy of an effect zone, two categories of personnel need to be considered: those persons who are in normally occupied buildings and those who are randomly located in the process areas performing operations and maintenance tasks. Estimating consequences to persons who are in normally occupied buildings is straightforward. If the normally occupied building is in an incident's effect zone, the occupants of the building are vulnerable to the effects of the incident. Vulnerability is defined as the probability that a receptor, whether human or equipment, will sustain a defined level of harm when exposed to the effect of an incident. For instance, if a person in a normally occupied building is exposed to a building collapse as the result of a vapor cloud explosion, they would have a fatal injury vulnerability of 0.6 according to estimates in *Guidelines for Evaluating Process Plant Buildings for External Explosions and Fire*.

For persons who are randomly located in the process area, the occupancy of the effect zone is a function of the population density. The population density is calculated by estimating the number of persons who are randomly located in the area of the process hazard at any given point in time. This number is estimated by dividing the relevant surrounding process area by the number of people expected to be located in that area.

Example 6.2

Problem: A process unit is 100 meters in length and 50 meters in width. The process unit has 10 operators and 10 maintenance technicians. The operators spend 10% of their time in the field and 90% in the control room. Maintenance staff on the other hand spend 40% of their time in the field, 20% of their time in the control room, and the balance in the maintenance shop. The staffing levels are constant regardless of shift. What is the occupancy of the control room? What is the transient population density in the field?

Solution: The control room is occupied 90% of the time by 10 operators and 20% of the time by 10 maintenance technicians.

Occupancy = $(10 \times 0.9) + (10 \times 0.2) = 11$ persons

The area of the process is 100 meters by 50 meters and is occupied 10% of the time by 10 operators and 40% of the time by 10 maintenance technicians.

Occupancy = $(10 \times 0.1) + (10 \times 0.4) = 5$ persons

Area = 100 meters $\times$ 50 meters = 5,000 meter^2

Density = 5 persons / 5,000 meter^2 = 0.001 persons/meter^2

Large impacts may require you to calculate different population densities for various zones. The population density for each zone inside a process plant is calculated as presented previously. If you expect offsite impacts, the calculation of population densities for these offsite zones will often have to be based on external databases containing land-use planning information.

The occupancy for an area occupied only randomly by people is a function of the size of the effect zone. The occupancy is determined by multiplying the effect zone area by the population density, as shown in equation 6.1:

$$O_{Person} = A_{Effect} \times \rho_{Person} \tag{6.1}$$

The consequence is then calculated by multiplying the occupancy by the vulnerability, as shown in equation 6.2. The consequence that results will depend on the effect that was analyzed. If the effect zone and corresponding vulnerability were determined for fatalities, then the consequence will be given in terms of probable loss of life (PLL).

$$PLL = O_{Person} \times V \tag{6.2}$$

Example 6.3

Problem: A toxic release has an effect zone of 950 ft^2, and is released into an area where the population density is 0.0003 person/ft^2 with no normally occupied buildings. The effect zone was determined for a 50% vulnerability of fatality. What is the consequence?

Solution:

PLL = (950 ft^2 × 0.0003 person/ft^2) × 0.5 = 0.14 persons for that individual event

We can estimate the financial consequence resulting from equipment damage in one of two ways. The most accurate method is to overlay the effect zone on a plot plan of the affected unit and tally all of the pieces of equipment that have been damaged. The equipment replacement component of the financial consequence will be the sum of the replacement costs of the damaged equipment. The replacement cost is determined by estimating the total amount of money that would be required to replace each individual process unit. The "book value," or amortized cost after depreciation, should not be used as this will yield a deceptively low cost. Estimation of the replacement cost of equipment items can be performed by using a company's internal cost estimation databases, vendor quotations, or reference books such as *Plant Design and Economics for Chemical Engineers* by Peters and Timmerhaus.

It is important to note that in many cases the financial cost of lost production is much greater than the cost of the individual pieces of equipment. This is especially true for the consequences of moderate equipment damage. Such equipment damage is small enough so that if the consequence analysis only considered this replacement cost, the corresponding safety system would not be very robust. However, the damage could easily be substantial enough that the production capacity of the plant is limited or shut off completely until replacement equipment can be made operational. If this equipment has any custom design features, the time required to return to full production can be significant. Thus, larger plants with process vulnerabilities of this nature must ensure that the cost of potential lost production is included as a primary part of the financial consequence analysis. This analysis requires very specific process expertise, so companies should designate it as a specific component of the overall analysis to ensure that the proper expert staff support is available.

A shortcut method for determining the financial consequences caused by equipment damage is to use equations 6.1 and 6.2, but replace personnel density with capital density. The capital density is thus the replacement cost of the process unit if it were to be destroyed divided by the process area. Capital density should be calculated for each individual process unit and not for the plant as a whole. Process unit replacement cost information can be obtained from the following sources:

- Estimates from site personnel.
- Purchase and installation costs scaled to current value using an appropriate cost index.
- Estimates based on general industry average data. Typical sources of information are *Handbook of Petroleum Refining Processes* and annual editions of *Hydrocarbon Processing* magazine including process descriptions.

6.6 Consequence Analysis Tools

A wide range of models is available for conducting consequence analysis. Table 6.1 presents some of the models most commonly used for consequence analysis as part of SIL selection. The table includes both publicly available and proprietary models used within industry and government and shows the strengths and limitations of the various models. Many models are commercially available; the table is by no means comprehensive. Many of the models listed were originally developed for analysis niches, such as EPA RMP compliance and environmental engineering. However, they can still be useful in more general applications.

An analyst should consider two different goals when determining which consequence modeling tools should apply to the SIL selection process:

1. Calculate a conservative estimate of the hazard distance with minimal effort.

   Selected SILs are order-of-magnitude assignments of SIS availability. Any methodology that creates results that are more accurate than order of magnitude at the expense of increased analysis time will not be cost-effective.

2. Calculate a very accurate description of the consequences of the release.

Example 6.4

Problem: A UOP Q-Max cumene production unit is built on a plot that is 200 meters in length and 100 meters in width. The loss prevention group estimates that an explosion could occur in the unit with an effect zone for 100% equipment damage vulnerability of 5,000 meter^2. What financial loss related to equipment damage can be expected from this incident?

Solution: The consequence analysis process begins by determining the capital density, which is the replacement cost of the unit divided by the unit's area.

Area = 200 meter × 100 meter = 20,000 meter^2

Hydrocarbon Processing magazine's 2001 “Petrochemical Processes” issue estimates that the erected cost of a UOP Q-Max process is approximately US$22 million.

Capital Density = $22,000,000 / 20,000 meter^2 = $1,100 / meter^2

According to the problem statement, the vulnerability to explosion of the equipment for the calculated effect zone is 100%. Since the effect zone and occupancy were calculated in terms of financial loss, the consequence will be expected value of loss (EV).

EV = $1,100 / meter^2 × 5,000 meter^2 × 100%= $5.5 million

Note that this analysis assumes a constant capital equipment density throughout the plot. This assumption may underestimate the damage that results from the most likely incidents since in most cases equipment is unevenly dispersed over a site. There is more likely to be an incident where there is more equipment, and thus the incident impact zone is more likely to contain that higher density of equipment and more-than-average damage will probably result.

You can avoid this pitfall of the simplified analysis by considering specific site plan information when conducting anything more than the most basic estimate.

If a simple consequence analysis tool produces results that are too conservative, the SIL you select will be unnecessarily high. Consequently, the installed SIS will be more costly than necessary.

Often these two goals are in conflict. A low-cost consequence analysis is usually very conservative and thus requires an expensive SIS to meet the conservative specification. On the other hand, the lower cost SIS often requires a very accurate (and thus expensive) consequence analysis to eliminate any conservatism in the system specification. Thus there is often a trade-off between the cost of the accurate consequence analysis and the cost of the safety system specified. For example, the FEMA's ARCHIE program (Automated Resource for Chemical Hazard Incident Evaluation) provides quick results, but these values are often very conservative, especially for toxic substances. Also, ARCHIE does not allow you to customize parameters such as weather conditions. A two-stage selection process should be considered when selecting an appropriate consequence analysis tool. As a first pass, perform a simple analysis. If this analysis yields unsatisfactory results from either a safety or cost perspective, then apply more advanced analysis.

Table 6.1 Comparison of Consequence Analysis Models

Model	Public/ Proprietary and Cost	Model Capability	Strengths	Limitations
FETCH™	proprietary developed by Exida low cost	• gas or liquid • buoyant or dense gas modeling • mixtures • explosion overpressure	• low cost • chemical database Web access	• gives very conservative results • limited flexibility
ARCHIE™	public; developed by EPA, FEMA, and DOT free	• gas or liquid • buoyant or dense gas modeling • mixtures • explosions	• openly available • credit for some passive mitigation (e.g., dikes)	• gives very conservative results for toxics • limited flexibility • no chemical database • DOS user interface
ALOHA™	public; developed by EPA and NOAA free	• gas or liquid • buoyant or dense gas modeling	• models hundreds of chemicals • easy to use • less conservative than OCA	• does not do fire or explosion overpressure calculation
DEGADIS™	public; co-funded by DOT, EPA, and DOE variable	• gas or liquid • dense gas modeling	• Windows™ - easy to use • chemicals can be pre-loaded • portions of model incorporated into ALOHA	• requires expert support • limited chemical database, can be supplemented
PHAST™	proprietary; developed by Det Norske Veritas high cost	• gas or liquid • buoyant or dense gas modeling • chemical database • mixtures • explosions	• DIPPR chemical database • can do aerosols • previous releases widely accepted within industry • good graphic ability	• dispersion may exceed EPA OCA • requires expert support
HGSYSTEM™	public; developed by joint industry / agency group free	• gas or liquid • buoyant or dense gas modeling • can do mixtures	• very good ammonia and hydrogen fluoride and some other dense gases • joint industry / government effort in validation	• very difficult to use; expert support needed • does not do overpressure • prone to input mistakes – DOS user interface • no graphics ability

- *Simple Analysis* - Using FETCH™, ALOHA™, or ARCHIE™ can be an efficient way to obtain a consequence estimate. Although the results may be extremely conservative in some cases, the relative ease of use and wide range of analyzed scenarios these models provide make them ideal for a first pass.
- *Advanced Analysis* - Using advanced proprietary or public domain models can be time-consuming and expensive. The important factors to consider in using a more advanced model are the relative difficulty of using the model, interpreting the results, and paying the cost to use the model in the first place. Also, the complexity of the tool itself may require that expert analysts execute the work. This combination will limit the practical use of advanced models to high-stakes situations.

6.7 Summary

Loss prevention practitioners have refined consequence analysis into a highly developed field. This chapter presented a variety of methods and tools available for performing consequence analysis, some of the reasons for conducting the analysis, and some of its potential pitfalls. We explained several of the terms used in the field in the context of the different elements of the analysis. We also reviewed several different consequences relevant to the process industry as well as the basic approach for translating the effect zones of an event into the magnitude of the harmful result as a whole.

Qualitative analysis methods are usually based on expert judgment and are generally best used for familiar systems in the relatively rare cases where few or no new features are present. Semi-quantitative methods are also available, but they are often difficult to harmonize with the other components of the analysis. Statistical analysis is a very effective tool for performing consequence analysis in certain situations, but unfortunately it has narrow application since it is limited by a lack of data applicable to the particular details of most process systems. Quantitative methods of varying degrees of rigor are thus the most useful consequence analysis tools and provide many options for the various situations an organization may face. The event tree classification method is especially useful for consequence analysis as part of the SIL selection process because it follows a form very similar to other SIL selection analyses.

Many process hazards are caused by the uncontrolled release of chemical potential energy. The specific outcome of this release depends on the properties of the chemical released and the circumstances surrounding the release. The chemical release event tree describes the specific outcomes possible for each type of hazard. Releases from chemicals hazards can result in flammable effects (i.e., flash fire, pool fire, jet fire, and fireball), explosion effects (i.e., vapor cloud explosion and physical explosion), and toxic effects. We can estimate the effect zones created by these incidents by using mathematical models that correlate the physical properties of the released material and the conditions of release to distances removed from the release where the effects are present.

Once you have established an incident's effect zone, you can determine the consequence of the release by estimating the occupancy of the effect zone. Consequences are typically determined in terms of fatality, injury, and financial loss caused by equipment damage and lost production. The consequence also depends on the occupant's vulnerability to the effects of an incident. Personnel occupancy can be determined in several ways, such as determining the occupancy of normally occupied buildings and determining the personnel density of randomly located workers in the process area. Financial losses due to equipment damage can be most accurately determined by identifying the specific items in an effect zone. They can also be estimated by using capital density. Lost production estimates typically require specific process expertise and can often dominate the financial components of the analysis.

In general, a large assortment of tools and techniques is available for quantitatively analyzing the potential impacts of chemical hazards. These tools range from screening index tools to simplified analysis tools to complex proprietary solutions. Our review of the commercially available consequence analysis tools in this chapter has shown that a two-tiered approach to model selection will provide reasonably accurate consequence analysis results without expending unnecessary resources. Thus, a good SIL selection protocol will include the use of simplified analysis tools to complete the bulk of consequence analysis scenarios. More advanced tools can then be used when these results are unsatisfactory, or when special conditions exist. A quantitative estimate of consequence, when combined with quantitative estimates of frequency, can effectively help an organization select the proper SIL for the situation at hand.

6.8 Exercises

6.1 List one semi-quantitative consequence estimation tool and explain why it is only semi-quantitative.

6.2 Explain the differences and similarities between physical explosions and vapor cloud explosions.

6.3 In the United States in 1998 there were 1,012 boiler explosions. In 1998, twelve fatalities resulted from boiler explosions. Estimate the consequence of a boiler explosion using statistical analysis.

6.4 An explosion has an effect zone with a distance from the center of the explosion to the edge of the effect zone of 25 meters. What is the effect zone area?

6.5 How can the release of a flammable gas result in a pool fire?

6.6 List three of the factors that determine the size of the effect zone for a toxic gas release.

6.7 Give an example of a consequence-modeling tool that performs computer calculations of toxic gas effect zones.

6.8 A process plant has a replacement cost of €25 million. The same plant has a plot plant 75 meters long and 84 meters wide. What is the average capital density of this plant?

6.9 The control room of a process is the work location for 20 engineers. These engineers spend about 5 percent of their time outside the control room. One plant supervisor is also in the control room 80 percent of the time. The 40 maintenance technicians that service the process spend about 10 percent of their time in this control room completing reports and assisting operators. What is the occupancy of the control room?

6.10 An explosion in the process area of a plant does not affect any normally occupied buildings. The personnel density is 0.002 persons per square meter, and the capital density is ¥150,000 per square meter. The explosion has a fatality effect zone of 5,600 m^2 and an equipment damage effect zone of 2,400 m^2. The vulnerability of both personnel and capital in these effect zones is 100 percent. What is the consequence of this explosion in terms of PLL and EV?

6.9 References

1. American Petroleum Institute. *API Recommended Practice 752 – Management of Hazards Associated with Location of Process Plant Buildings.* Washington, DC: American Petroleum Institute, 1995.

2. Center for Chemical Process Safety. *Guidelines for Chemical Process Quantitative Risk Analysis.* New York: American Institute of Chemical Engineers Center for Chemical Process Safety, 2000.

3. Center for Chemical Process Safety. *Guidelines for Consequence Analysis of Chemical Releases.* New York: American Institute of Chemical Engineers Center for Chemical Process Safety, 1999.

4. Center for Chemical Process Safety. *Guidelines for Evaluating Process Plant Buildings for External Explosions and Fire.* New York: American

Institute of Chemical Engineers Center for Chemical Process Safety, 1996.

5. Environmental Protection Agency. *Risk Management Program Guidance for Offsite Consequence Analysis.* Washington, DC: United States Environmental Protection Agency, 1999.

6. Federal Emergency Management Agency, Department of Transportation, Environmental Protection Agency. *Handbook of Chemical Hazard Analysis Procedures.* Washington, DC: Federal Emergency Management Agency, 1990.

7. Glasstone, S., and P. J. Dolan. *The Effects of Nuclear Weapons,* 3d ed. Tonbridge Wells, UK: Castle House Publishers, 1980.

8. Lees, F. P. *Loss Prevention for the Process Industries.* London: Butterworth and Heinemann, 1992.

9. Meyers, Robert E. *Handbook of Petroleum Refining Processes,* 2d ed. New York: McGraw-Hill, 1997.

10. Mudan, K. S., and P. Croce. *NFPA/SFPE Handbook of Fire Protection Engineering,* section 2, chapter 4. Batterymarch, MA: National Fire Protection Association, 1996.

11. Peters, Max S., and Klaus D. Timmerhaus. *Plant Design and Economics for Chemical Engineers,* 4th ed. New York: McGraw-Hill, 1991.

CHAPTER 7

Likelihood Analysis Overview

Understanding likelihood analysis, like consequence analysis, is an important part of understanding the overall risk of a process. As we noted earlier, a knowledge of the magnitude of consequences alone provides an incomplete understanding of risk. You can determine the remaining likelihood component of risk using a variety of methods, from qualitative study to statistical analysis to investigation using several fault propagation models.

We usually determine failure rates and event frequencies through historical experience. For complex situations, such as the failure of a process plant, there is not enough sufficiently relevant historical data to directly form valid statistical conclusions. In these situations, we determine failure rates by breaking the overall failure into smaller, more quantifiable parts through a method called fault propagation modeling. Unlike the relatively unique characteristics of the process as a whole, these process event segments are often similar to other small process plant event segments. They can therefore be compared statistically to predict their individual likelihood or probability. We can then combine these small event probabilities to calculate the likelihood that the overall sequence will lead to a harmful failure of the larger system in question.

Fault propagation modeling is the process of determining the frequency or probability of a complex sequence of events (leading to an accident) based on the frequencies of the events that initiate or contribute to the resulting accident. Several different fault propagation modeling techniques are commonly used. The best model to use for a particular situation will depend on the complexity of the scenario under study and the accuracy of the results required. The primary models used today include fault tree analysis, event tree modeling, block diagrams, and Markov analysis, all of which we will discuss in this chapter.

7.1 Statistical Analysis

The likelihood that smaller component events could occur, and thus contribute to an overall failure, is often determined by statistically analyzing historical data. This is performed by dividing the number of failures that have been recorded for the process system by the amount of time that the system was in operation, as described by equation 5.9. This procedure has been used to tabulate the failure rates of systems ranging from the very simple, such as a control valve, to the relatively complex, such as an industrial boiler. General data on failure frequencies for a wide range of common systems and subsystems can be found in industry reference books. Also, many companies that perform risk management services, or provide insurance, maintain databases of this information.

Some examples of the event likelihoods that are available in general reference books include the likelihood in a given year of death by lightning strike, which is 1×10^{-7} (Smith), and the annual likelihood of an explosion in a Catalytic Reforming Unit, which is 2.6×10^{-4} (API). There are a large number of reference sources that contain statistical accident data for a variety of industries.

Unfortunately, statistical analysis cannot always be directly applied to the determination of the failure rates of process plants as a whole. Just as with the statistical analysis of consequences, two conditions must be met before we can make a valid statistical inference. First, there must be a sufficiently large amount of data to be analyzed. Second, all of the data must come from systems that are roughly similar so that any conclusions will be valid for the case in question. These two conditions tend to conflict and are rarely met when analysts consider the overall system-level event frequencies for process plants.

For example, if an engineer has fifteen years of experience with a full crude unit system, what explosion likelihood can he or she estimate? If no events have ever occurred with the specific unit, any assumption the analyst makes that an event will or will not occur can have no longer basis than the test period of his or her own experience. In this case, the analyst must assign the explosion likelihood as no less than once in fifteen years or $\frac{1}{15}$ annually. When trying to select SILs for process plants, one will often need to show that accidents will only occur within a period of 10,000 to 100,000 years! Ideally, one would need data from other identical units, but again, very few other units are similar enough from a risk analysis standpoint to justify combining data sets. These other units, in different plants, will each have varying throughputs, operating conditions, equipment types, and staffing philosophies. Each of these differences will make failure event frequencies drastically different, and thus typically impossible to apply to the situation at hand in a direct one-to-one fashion.

7.2 Fault Propagation Modeling

When statistical analysis alone is not an effective way to determine event frequency, *fault propagation modeling* can be used. Fault propagation modeling is the analysis of the chain of events that leads to an accident. By analyzing what events initiate that chain and which events contribute to or allow the accident to propagate, and then by establishing how these events are logically related, you can determine the final event frequency. Fault propagation modeling techniques use the failure rates of individual components to determine the failure rate of the overall system. As we have seen, the failure rates of individual components are easier to find in reference databases or to estimate directly, because the component is usually simple and is readily comparable to other components in similar types of service. The failure rates of individual components such as pumps, compressors, valves, and vessels are easier to find in reference databases or to estimate because a large experience base is available for them. For example, although there may only be one crude unit at a facility, there may be several hundred pumps in the process plant as a whole. With data from several hundred pumps, developing thousands of years of operating data is much easier. Also, the differences between one pump and another in a single facility are often not as extreme as the differences between one process plant and another.

Fault propagation modeling uses probability math to combine the failures of the individual components. Depending on the type of model used, different logical operations are allowed and thus different types of scenarios can be analyzed. As we mentioned, several fault propagation modeling techniques are in general use today, including event tree analysis, reliability block diagrams, fault tree analysis, and Markov analysis.

7.2.1 Event Tree Analysis

The most straightforward and easy-to-use modeling technique is *event tree analysis*. An event tree begins with an initiating event, usually displayed on the left of the event tree diagram (see figure 7.1). The tree expands from the initiating event in branches representing possible subsequent outcomes. Each branch traces a situation in which a different outcome is possible. The event tree ends with multiple possible outcomes. Although event trees are straightforward, they have limited application because they cannot analyze complex situations. We will describe event trees in much more detail in chapter 8.

7.2.2 Reliability Block Diagrams

The *reliability block diagram* is a more robust likelihood-modeling tool that is still relatively easy to use. In reliability block diagrams, a block represents each equipment item in a system, and the arrangement of the blocks represents the logical relations between the potential failures of the components (see figure 7.2). This model can then be used to calculate the probability of failure of the overall system. Items that are vertically aligned represent a parallel path (OR logic), and items that are horizontally aligned represent a series path (AND logic). Some blocks represent complex equipment groups such as two-out-of-three voting. The failure

Figure 7.1 Typical Event Tree

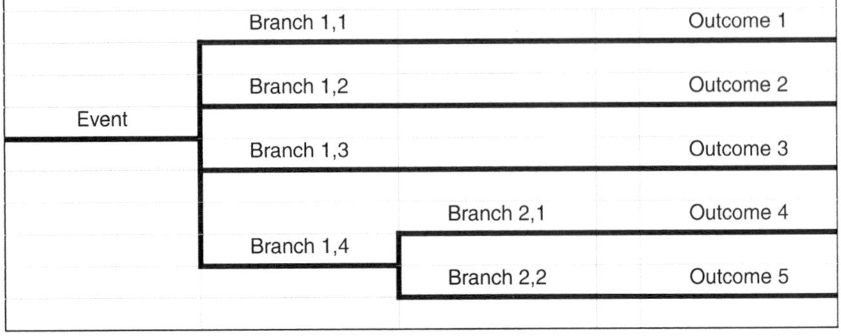

Event
Branch 1,1
Branch 1,2
Branch 1,3
Branch 1,4
Branch 2,1
Branch 2,2
Outcome 1
Outcome 2
Outcome 3
Outcome 4
Outcome 5

rate of these blocks is calculated using special equations that have been derived for each case. This method has the strength of being easy to use, but it is limited in practice to situations in which the governing equations have already been derived.

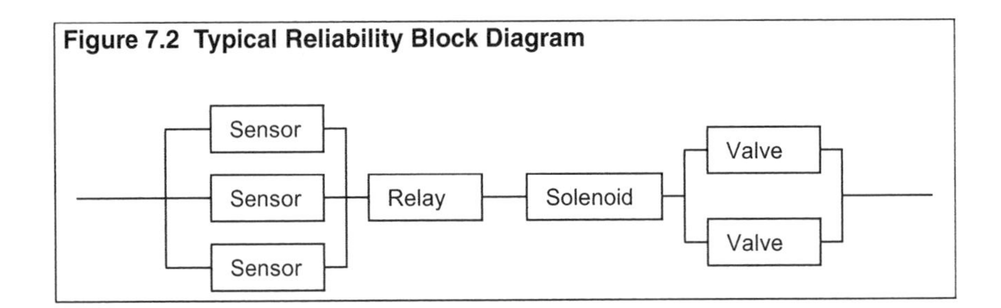

Figure 7.2 Typical Reliability Block Diagram

7.2.3 Fault Tree Analysis

Fault tree analysis can address more complex logic than either event tree analysis or reliability block diagrams, but it too is still limited by the need to use prederived equations to calculate the failure rates of basic events. Fault tree analysis is also weak when a system's performance must be analyzed over time. As with the other methods, a diagram is constructed that represents the logical relationships between initiating, intermediate, and top events (see figure 7.3). The objective of fault tree analysis is to determine the likelihood of the top event, which is generally chosen to be an overall system failure or harmful accident outcome. The top-event frequency is a function of the basic events that are located at the bottom, or ends, of the fault tree branches. The basic events are connected to the top event through a series of logic gates. Gates include OR gates, which perform probability addition of the inputs, and AND gates, which perform probability multiplication of the inputs. Once the model is constructed, the top-event probability is calculated using known probability formulas. Fault trees are described in more detail in section 5.4.

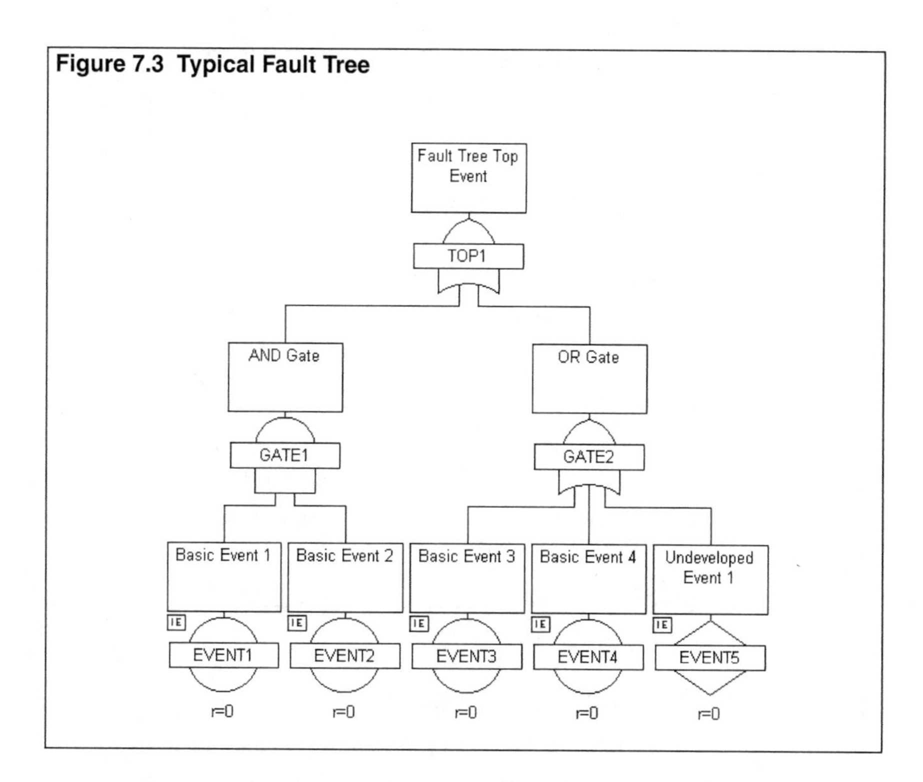

Figure 7.3 Typical Fault Tree

7.2.4 Markov Analysis

Markov analysis is the most complex of the methods mentioned here; it is also the most accurate and flexible. This method is often used for very complex systems such as fault-tolerant programmable logic controllers (PLCs). As with the other modeling tools, a diagram is constructed to represent the system, including the logical relationships between its components. In Markov analysis there is a group of circles, each of which represents a system state. The Markov model shown in figure 7.4 represents the failure states and transitions between failures for a "1oo1" (one-out-of-one) voting subsystem. This means that there is one element present, and its vote alone can shut down the system. For the 1oo1 system, there are four states: OK (0), Failed Safe (1), Failed Dangerously but Detected (2), and Failed Dangerously Undetected (3). The different states are connected by transitions, which are shown as arrows and indicate paths for moving from one state to another. These transitions are quantified by using either failure rates (λ), when the transition is from an OK state to a failed state, or repair rates (μ) when the transition is from a failed state back to an OK state. As with the other fault-propagation models, the diagram is converted into a set of mathematical equations that the analyst then solves to quantify the likelihood of the event or state in question. W. M. Goble presents an excellent discussion of Markov analysis in his *Control Systems: Safety, Evaluation, and Reliability.*

Figure 7.4 Typical Markov Model

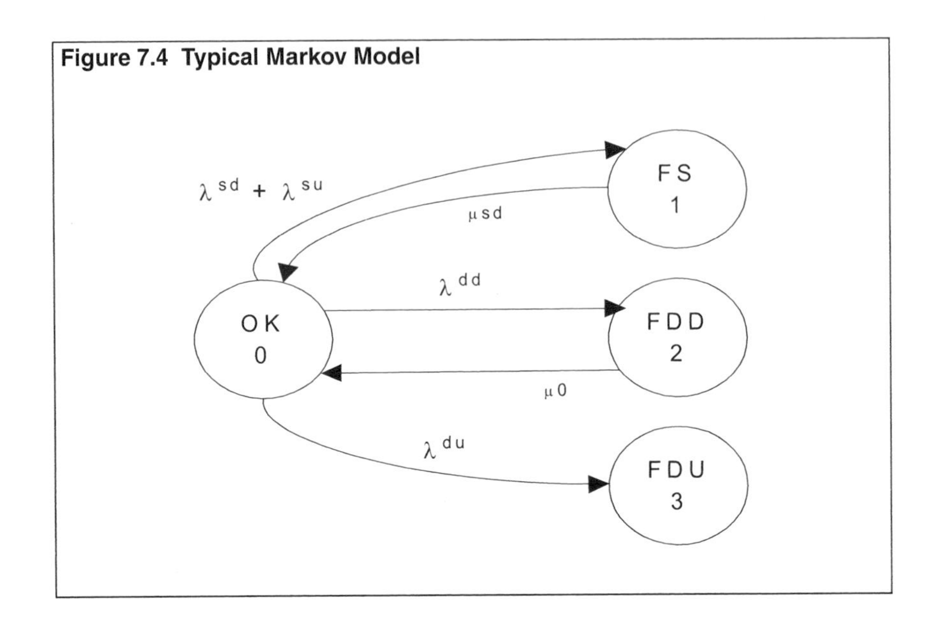

7.3 Likelihood Analysis: An Example

In this section we will solve a sample problem using each of the four techniques to illustrate the differences between fault propagation modeling techniques and to develop a better appreciation for when the various fault propagation models are most appropriate. The example we will consider is a likelihood analysis where the goal is to estimate the likelihood of a process accident as part of the overall SIL selection process. As the sample problem is developed, it will become apparent that event tree and fault tree tools lend themselves more readily to the problem of analyzing the likelihood of process accidents. It will also become clear that reliability block diagrams and Markov models are better suited to analyzing mechanical systems without manual intervention.

7.3.1 Problem Statement

A safety instrumented function (SIF) is being considered to protect piping and vessels downstream of a pressure-reducing station (see figure 7.5). Natural gas from a pipeline enters the system at point a, at 4,000 kPa(g). The pressure is reduced to 900 kPa(g) as measured by PT-01, located downstream of the choke valve, PV-01, and prior to entering the liquid knockout drum. Piping and vessels downstream of the choke valve are designed for a maximum pressure of 1,500 kPa(g).

A hazard assessment has determined that if the pressure-reduction control system fails so that the choke valve (PV-01) is in the open position, then the liquid knockout drum may rupture and cause a release of high-pressure natural gas. This could potentially cause a fireball and subsequent jet fire. The loss prevention team analyzed the failure scenario

Figure 7.5 Example of Pressure-Reducing Station

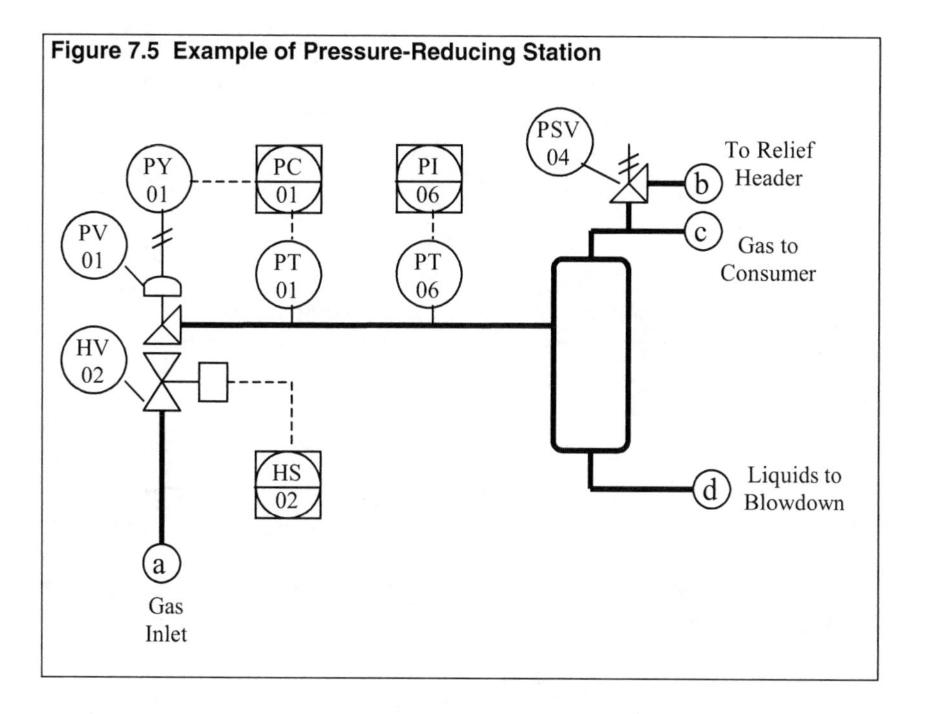

and determined that developing enough pressure to rupture the vessel would require at least 25 minutes because of the large inventory of the pipeline leading to and from the knockout drum and based on the design of the choke valve. A historical reliability analysis of this loop and hundreds of similar ones owned by the same company yielded a failure rate in the open-valve position of once in twenty years.

If the automatic control system fails, there are two existing safeguards that can prevent propagation into a vessel rupture. First, the liquid knockout drum is equipped with a pressure relief valve that has been designed for this scenario. The plant's loss prevention team has determined that the probability of failure for this type of relief valve in natural gas services is 0.05. The second protection layer is operator intervention. If the operator detects a high pressure from an alarm generated by the independent redundant pressure transmitter PT-06, he or she can activate HS-02, closing a shutoff valve upstream of the choke valve. The loss prevention team has determined that human error is the major contributor to the failure of this protection layer. A simplified analysis yielded a human error probability of 0.1 for this scenario.

Based on this information, what is the rate at which a vessel rupture incident will occur without the addition of an SIF?

7.3.2 Event Tree Solution

The event tree solution begins by building the graphical description of the logic that relates the initiating event to the potential outcomes (see figure 7.6). The initiating event in this problem is the failure of control loop PC-01. This event is shown as a line coming in from the left side of the tree. The two protection layers that are available form the two branches of the tree. Each branch has two options, either the protection layer operates or it fails. As a result, three outcomes are possible: vessel rupture, system shutdown by operator intervention, or system vented through relief valve.

Figure 7.6 Solution to Example Using Event Tree Analysis

Pressure Control Loop Fails Valve to Open (per year)	Operator Diagnoses Problem and Closes Valve	Relief Valve Opens and Vents System	Outcomes
		0.05	2.50E-04
		Failure	Vessel Rupture
	0.1		
	Failure	0.95	4.75E-03
0.05		Success	Relief Valve Vents System
PC-01 Fails			
	0.9		4.50E-02
	Success		System Shutdown by Op.

The frequencies of each of the outcomes are calculated using probability multiplication. For each outcome, the frequency is the initiating event frequency (i.e., the frequency at which PC-01 fails) multiplied by all of the probabilities on the path to that outcome. For the case of the vessel rupture, which is the only one we are concerned with, the frequency of the outcome is the frequency of the initiating event multiplied by the probability of failure of both protection layers, or 2.5×10^{-4} per year.

As this example demonstrates, building an event tree is straightforward, and the computational effort required to obtain a result is minimal. In chapter 9, event trees will be simplified even further into layer of protection analysis (LOPA).

7.3.3 Fault Tree Solution

As with the event tree solution, the fault tree solution begins with the development of a graphical representation of the logic that relates the basic events. In this case, the basic events are PC-01 failure, operator intervention failure, and relief valve failure. The top event of the fault tree is a vessel rupture. A fault tree that describes this scenario is shown in figure 7.7.

Logically, a vessel rupture occurs when the initiating event, "PC-01 Fails," occurs AND the safeguards fail. In this case, the safeguards fail if both the relief valve fails AND operator intervention fails. This fault tree,

Figure 7.7 Solution to Example Using Fault Tree Analysis

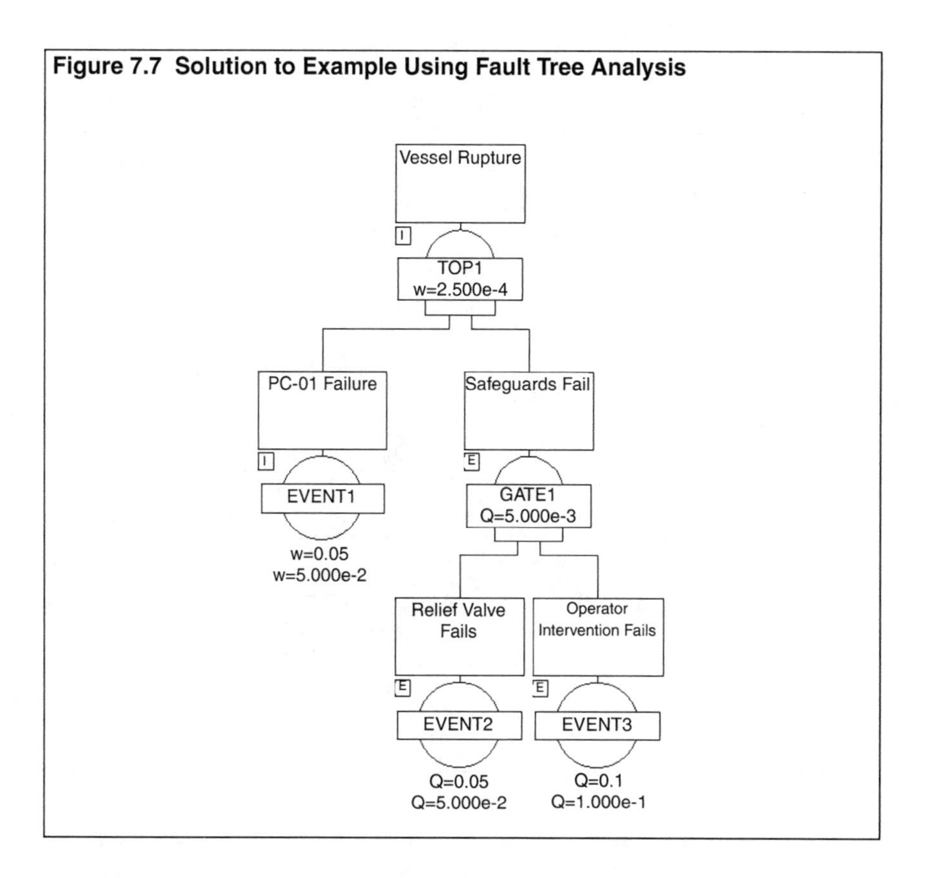

shown in figure 7.7, is typical of a fault tree that is equivalent to a LOPA in that there is only one initiating event, as we will see in chapter 9. We could have obtained the same result by using a fault tree in which all three basic events are connected to the top event under an AND gate. In either case, the result is effectively obtained by multiplying the initiating event frequency by all of the protection layer failure probabilities. As with the event tree solution, the fault tree solution yields a 2.5×10^{-4} per year frequency.

7.3.4 Reliability Block Diagram Solution

Using reliability block diagrams to model the failures of process systems is somewhat more difficult than with the two previous methods. There are two reasons for this. First, this method is designed to calculate system success instead of system failure. As a result, you will have to convert all failure rates and probabilities to success and then convert the results back from success to failure. Second, as with fault trees, the reliability block diagram method cannot directly employ frequencies. Instead, users are required to convert initiating event frequencies into probability over a given time interval, and then convert the calculated failure probability over that time interval back into a frequency.

We begin the solution by developing a diagram that represents the way in which the components of the system are logically related. The system is successfully operational if either the controller is operating, the operator is able to manually vent the system if a failure occurs, or the relief valve is able to vent the system if a failure occurs. Since the system is successful if any of the individual components are successful, we draw the diagram as three boxes connected in parallel. The reliability block diagram that describes this situation is shown in figure 7.8.

Figure 7.8 Solution to Example Using Reliability Block Diagrams

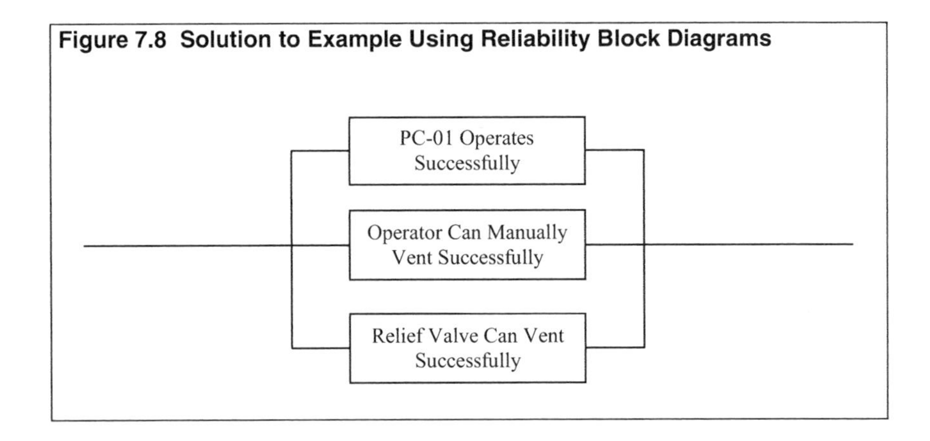

In reliability block diagrams, we perform the solution of a parallel network by a non-mutually exclusive probability addition, as shown in equation 5.7. Before performing this probability addition, all components must be described in terms of probability of success. For manual intervention and relief valve success, this is performed using equation 5.3 for complementary events. Therefore:

$$P_{S,RV} = 1 - 0.05 = 0.95$$

$$P_{S,Manual} = 1 - 0.1 = 0.9$$

Obtaining the probability of the successful operation of PC-01 is more complex. We will arbitrarily define a one-year time interval for which we wish to know the probability of success. We can then calculate this probability using a rearrangement of equation 5.11 in which success probability is determined instead of failure probability:

$$P_{S,PC} = e^{-((1/20) \times 1)} = 0.951$$

We then combine the success probabilities we have just calculated using equation 5.7 to obtain an overall system success probability:

$$P_{S,SYSTEM} = 1 - [(1 - 0.95) \times (1 - 0.9) \times (1 - 0.951)] = 0.999755$$

Then the probability of failure is calculated using equation 5.3:

$$P_{F,SYSTEM} = 1 - 0.999755 = 2.45 \times 10^{-4}$$

This result represents the probability that the system will fail (i.e., a vessel rupture will occur) over a one-year period. Converting this probability back into a frequency again requires a rearrangement of equation 5.11:

$$F = (-\ln(1 - 2.45 \times 10^{-4})) / 1 = 2.45 \times 10^{-4}$$

The resulting frequency of vessel rupture is 2.45×10^{-4}. The reliability block diagram, as expected, yielded the same result as the two previous methods, but a lot more work was required because this method is not well suited to the problem at hand.

7.3.5 Markov Model Solution

The Markov method requires considerably more effort than the previous methods to develop the framework necessary to arrive at the solution. However, once we have developed the framework, we can use it to provide much richer information about the characteristics of the system. The difficulties it poses relate to the time-dependent perspective Markov analysis takes, as well as to the difficulty of expressing each of the potential contributing component failures in that analysis. A full discussion of these issues and the background required to fully address them are beyond the scope of this text, so we again refer the reader to the presentation by Goble in *Control Systems: Safety, Evaluation, and Reliability.*

Once you have gone through the somewhat tedious exercise of building a model that represents all potential system states, quantifying the probabilities of moving between states, building a transition matrix using those probabilities, and justifying the appropriate assignments and time periods represented in figure 7.9, the final matrix multiplication step is relatively straightforward. This entire process gives the probability of being in the overpressure event state during the reference time period as 2.5×10^{-4}. As one would hope, this is consistent with the results of the other methods.

7.3.6 Comparing the Solution Techniques

We have seen that several solution techniques are available for performing likelihood analysis, each with its own strengths and limitations. While the event tree method is considered inflexible, it is well suited to the analysis that is typically required to determine the likelihood of process accidents caused by initiating events and the failure of safeguards. Its simplicity, ease of use, and inherent ease of documentation make it ideal for this application.

Figure 7.9 Solution to Example Using Markov Model

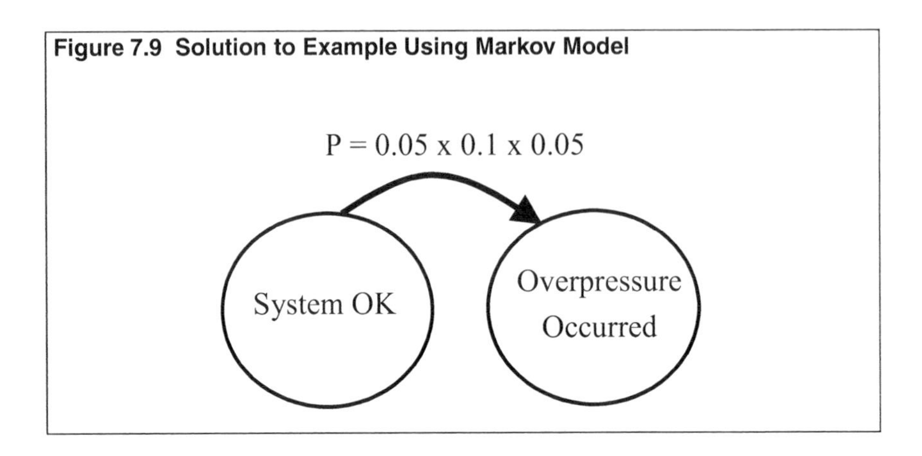

7.4 Summary

This chapter described some of the basic methods for likelihood analysis. These methods cover an extremely diverse range, from simple qualitative to statistical methods to fault propagation modeling. Each method has strengths and limitations so the proper method for a given situation will depend on the details of that particular case.

Statistical analysis is used to determine failure rates for a wide range of events. Failure rate data for processes and equipment is available in standard industry references. Although statistical analysis is quite appropriate for determining the failure rates of equipment items such as pumps and valves, it is not usually valid for estimating the rate at which complex systems fail. There is typically not enough operating experience or similarity between processes to make valid statistical inferences about complex systems. However, statistical modeling is useful for determining the component failure rates for the individual events or component failures that can combine and lead to full system failures. These component failure rates serve as input for the various techniques in the fault propagation family also presented here.

These fault propagation models can be extremely useful for calculating the failure rates of complex systems. Fault propagation modeling allows an analyst to determine an event rate based on the rates of initiating and contributing events and on the ways those events are logically related to the system failure in question. There are several types of fault propagation models in general use, including event trees, fault trees, reliability block diagrams, and Markov models. All are based on the idea of constructing a diagrammatic representation of the events and interconnecting logic that lead to system failure. These diagrams then allow you to apply mathematical equations to the systems to determine the ultimate failure rate or likelihood in question.

7.5 Exercises

7.1 Compare and contrast the primary methods for fault propagation modeling.

7.2 When is fault propagation modeling preferred over statistical methods for process plant failures?

7.6 References

1. American Petroleum Institute. *API Recommended Practice 752 – Management of Hazards Associated with Location of Process Plant Buildings.* Washington, DC: American Petroleum Institute, 1995.

2. Goble, William M. *Control Systems: Safety, Evaluation, and Reliability.* Research Triangle Park, NC: ISA, 1998.

3. Lees, F. P. *Loss Prevention for the Process Industries.* London: Butterworth and Heinemann, 1992.

4. Smith, D. J. *Reliability, Maintainability, and Risk.* London: Butterworth and Heinemann, 1993.

CHAPTER 8

Event Tree Analysis

Event tree analysis is the most straightforward of the common fault propagation modeling techniques. This chapter describes how to use event tree analysis to both determine the frequencies of events and to average consequences where multiple outcomes are possible from a single incident. Both the physical appearance of the tree and the logical relationship between the events are explained. The basic tree begins with an initiating event that starts one or more chains of intermediate events leading to their various final outcomes. These trees will have one or more branches of multiple event possibilities that determine which outcomes can result. The outcomes are then at the end of each branch, and their frequency or probability is calculated by following the sequence set out by the branched paths that lead from the initiating event.

Solving for event tree outcome frequencies or probabilities is an exercise in probability math. Each individual outcome's probability or frequency is a function of the likelihood of each of the events along the path from the initiating event to that outcome. When using event trees to average outcome consequences, you must consider the probability of each branch together with its consequence. This probability acts to weight or adjust the different outcomes, which you then total to get an overall result.

8.1 Introduction to Event Tree Analysis

As we discussed in the previous chapter, event tree analysis is a form of fault propagation modeling. This particular method is well suited to estimating the risk stemming from process plant failures. Neither the initiators of industrial accidents nor the layers of protection that prevent them are typically complex, that is, they do not usually require analysis by redundant systems or the time for on-line repair. Thus, they can be accurately characterized with the probability multiplication methods that form the basis of event tree analysis. In addition, the diagrams produced during the analysis are descriptive of the scenario and clearly convey the key features, which makes this method a good way to document the risk analysis process.

A typical event tree is shown in figure 8.1. An event tree begins with a single initiating event, which is usually an action or a failure of a piece of equipment that starts the chain of events leading to one of several outcomes. The event tree's branches determine the different event sequence paths to the various outcomes. Each branch consists of a set of events that can occur in the chain leading to the final outcomes. Each branch associates a probability of occurrence with each possible event in the branch set.

Figure 8.1 Typical Event Tree

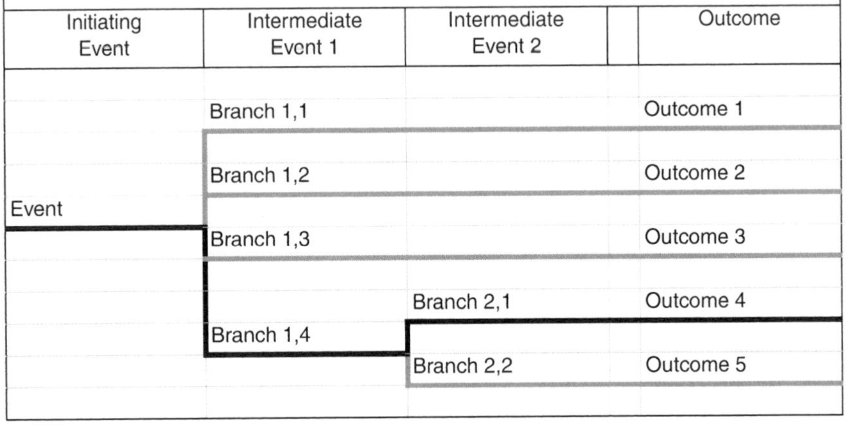

Initiating Event	Intermediate Event 1	Intermediate Event 2	Outcome
	Branch 1,1		Outcome 1
	Branch 1,2		Outcome 2
Event	Branch 1,3		Outcome 3
		Branch 2,1	Outcome 4
	Branch 1,4		
		Branch 2,2	Outcome 5

With all of these different multi-event branches, it is typical for an event tree to have multiple outcomes. The probability of each outcome is a function of the path to the outcome and the probability of each event along that path. The probability of each outcome is calculated using probability multiplication. For an outcome to occur, the initiating event must occur in combination with all of the branch events that connect the initiating event and the final outcome. The relation between the initiating events and event tree branch probabilities is a logical 'AND'. For instance, figure 8.1 shows the path to outcome 4 in bold black. In order for outcome 4 to occur, the following events must also occur: the initiating event, option 4 from the intermediate event 1 set, and option 1 from the intermediate event 2 set.

8.2 Initiating Events

An initiating event starts the chain of events that can lead to the unwanted accident if one of the protection layers does not prevent it from propagating. Any number of things can initiate an unwanted accident, from the failure of a piece of process equipment or instrumentation, such as a pump failing to provide cooling water, to the erroneous action of an operator, such as mistakenly setting the cooling water flow control set point too low. Some initiating events are not accidental. For instance, a batch process might have the potential to undergo a temperature runaway and explode. In this situation, the initiating event is the loading or mixing of the reactive materials at the beginning of the batch run.

Initiating events are usually quantified by their frequency of occurrence. For instance, a cooling water pump might fail once every 5.7 years, yielding an annual frequency of 1/5.7. Sometimes, event trees are completely described using probabilities; in these cases the initiating event will also be expressed as a probability. That probability will then be based on either a certain number of specific opportunities that the initiating event could occur or else based on a specified time period.

As we discussed in chapter 7, identifying the frequency or probability of an initiating event is a critical component of an event tree analysis. Every effort should be made to establish that value as accurately as possible using the most relevant information available.

8.3 Branches

The branches of an event tree are groups of events where different outcomes are possible depending on which event or events of the set is true. The intermediate outcome selected from each set of events that make up each branch determines the overall outcome.

Branches of event trees are usually complementary events. For instance, a branch event could be the failure of a relief valve. The event set includes two complementary events, namely: (1) relief valve fails and (2) relief valve operates. Although complementary events occur often, this is not always the case. For instance, an event tree branch might be the state of the chemical that is released, with a set of three possible states: (1) gas, (2) liquid, and (3) solid. In this case, the set of events does not even have to be mutually exclusive. If liquefied petroleum gas, or LPG, were released, the state of the release would be liquid and gas, so both of the branches would be true.

Figure 8.2 shows a branch that addresses the issue of the state of a material from a perspective other than simply its state during the release. Here, the initiating event is such that material is always released as a liquid. However, the external conditions may cause a phase change sufficient to create a flammable vapor cloud. In addition, the branch provides a choice of possible magnitudes for the release. This makes it possible to differentiate the eventual outcomes of small fires that can be readily managed with negligible consequence and large fires that require significant external assistance and can cause major harm. Figure 8.2 is clearly incomplete since at least a second set of branch events will be required to address the potential that the flammable clouds and/or liquid pools will be ignited.

Each of the events in each branch of an event tree has a probability of occurrence associated with it. When performing event tree analysis, you

Figure 8.2 Branches in an Event Tree

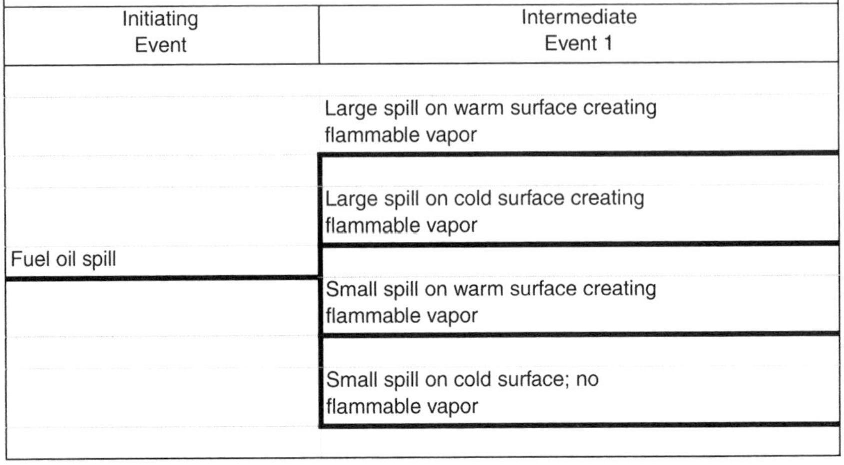

Initiating Event	Intermediate Event 1
	Large spill on warm surface creating flammable vapor
	Large spill on cold surface creating flammable vapor
Fuel oil spill	Small spill on warm surface creating flammable vapor
	Small spill on cold surface; no flammable vapor

must use probability, not frequency, in the branches. While multiplying frequency by probability yields a valid result, multiplying two frequencies together does not. If the effectiveness of a safeguard is presented in terms of failure rate, you will need to convert it into a probability over a certain reference time period or number of occasions before using it in an event tree.

8.4 Outcomes

Event tree analysis always results in multiple outcomes. Although only one outcome may be of interest, the others are also presented and can be calculated. The number of possible outcomes is a function of the number of events in each branch.

The outcomes of an event tree are typically described in terms of frequency, although under certain circumstances outcome probabilities will be desired. The outcome probability form is generated by describing the initiating event in terms of probability. Calculating the average consequence of an outcome is a good example of an occasion when event trees are quantified using probabilities instead of frequencies. We explain this instance in detail in section 8.6.

The path from an initiating event to an outcome defines the series of intermediate events that must occur for that outcome to result. It also defines the logic that must be used to show how these events are related. When following a path from the initiating event to an outcome, we relate the initiating event and branch events to the outcome by using a logical "AND." For an outcome to be true, the initiating event must be true as well as all of the branch events that lead to the outcome.

Example 8.1

Problem: Draw an event tree for getting a flat tire. Assume that the initiating event is running over a nail. Also assume that the nail may not puncture the tire, and even if the tire is punctured, it may not deflate.

Solution: The problem states that the initiating event for the tree is running over a nail. The problem also states that there are two branches for the tree. The first branch is the possibility that the nail may or may not puncture the tire. The second branch is the possibility that the tire may or may not deflate after being punctured.

When drawing the tree, the beginning point is the initiating event, usually positioned on the left. Next, from the initiating event, draw each branch. In this example there are two branch event sets. The first branch is the possibility that the nail punctures the tire. There are two possible events in this set, thus two branch paths. The second branch set is the possibility that the puncture does not cause the tire to deflate. The second branch only impacts one of the events that is possible in the first branch set, since the puncture must have occurred for this situation to be considered. The order in which the branches are shown on the tree is important in some cases, and should be considered when building an event tree.

INIT EVENT	BRANCH 1	BRANCH 2	OUTCOME
Run over nail	Nail punctures tire	Tire deflates due to puncture	
		TRUE	Tire Deflates
	TRUE		
		FALSE	Puncture, no deflation
	FALSE		No Puncture

Three outcomes are possible as a result of this initiating event: (1) the tire deflates, (2) the tire is punctured but does not deflate, and (3) the tire is not punctured.

8.5 Quantifying Event Trees

The probability or frequency of an event tree outcome is calculated as the logical combination of the events that fit together to cause the outcome. This combination includes the initiating event and the intermediate branch events in the path from the initiating event to the outcome. As stated earlier, these events are related through a logical "AND." In order for an outcome to be true, the initiating event must be true AND the first branch event must be true, and so on.

When events are related by a logical AND, the probability of the combination of events is calculated using probability multiplication. (See section 5.3.1 for more information on probability multiplication.) In other words, the outcome frequency is the initiating event frequency multiplied by the probabilities of each of the branch events on the path to the outcome.

Example 8.2

Problem: In example 8.1, an event tree diagram was built to describe the events that might result from running over a nail with a car tire. Quantify this event tree, including all possible outcomes, using the following data:

Running over a nail	once in 0.25 years
Nail punctures tire	once in 50 attempts
Punctured tire deflates	once in 5 events

Solution: Add the frequency and probability data to the event tree as shown in the following diagram. Note that the probabilities were added to both of the events in both branches. This was done by assuming that the events were complementary. Calculate the final outcome probabilities by using probability multiplication.

For the outcome of Tire Deflates calculate the frequency as follows:

$$(1/0.25) \times (1/50) \times (1/5) = 0.016 \text{ /year}$$

For the outcome of Puncture, but No Deflation, calculate the frequency as follows:

$$(1/0.25) \times (1/50) \times (4/5) = 0.064 \text{ /year}$$

For the outcome of No Puncture, calculate the frequency as follows:

$$(1/0.25) \times (49/50) = 3.92 \text{ /year}$$

INIT EVENT	BRANCH 1	BRANCH 2	OUTCOME
Run over nail	Nail punctures tire	Tire deflates due to puncture	
		1/5	0.016/yr
		TRUE	Tire Deflates
	1/50		
1/0.25 years	TRUE	4/5	0.064/yr
		FALSE	Puncture, No Deflation
	49/50		3.92/yr
	FALSE		No Puncture

8.6 Average Consequence of Incidents Using Event Trees

Event trees are used extensively to determine outcome frequencies. They are also well suited to assisting in the calculation of an average consequence. As chapter 6 demonstrated, a single incident can have multiple incident outcomes. For instance, the release of a flammable gas can result in vapor cloud explosion, flash fire, jet fire, or even have no significant impact at all. The factors that determine which outcome will occur include probability of ignition, probability of explosion, and the conditions surrounding the release. Chapter 6 also demonstrated the range of potential incident outcomes using event trees.

Each of the possible incident outcomes will have a different consequence. A flash fire will often have a much smaller consequence in terms of equipment damage than a vapor cloud explosion that has the same flammable mass and energy release potential. In addition, if no source of ignition is present in the flammable zone before the cloud disperses, the consequence may be insignificant. We can combine all of these considerations using an event tree whose outcomes are quantified in terms of probability rather than frequency.

To determine an average consequence of an incident, draw the event tree with the incident as the initiating event. Then draw the event tree with branches that will determine the various possible outcomes, such as probability of ignition. With this method, the resulting event tree outcomes will correspond to all of the potential incident outcomes. You should quantify the consequence of each of the outcomes in terms of the impact being measured. As we noted in chapters 3 and 6, these terms can include probable loss of life (PLL), probable injuries, financial loss, or any other appropriate measure of consequence.

With the average-consequence-of-incident method, the initiating event is simply quantified by stating that the incident is 100 percent likely to occur. A probability of 1 is therefore assigned to the initiating event to properly normalize the calculation. Using branch event probabilities, we can calculate the probability of each of the outcomes as for a normal event tree. We can then determine the average consequence by weighting each potential incident outcome by the probability that we calculated using the event tree, and then summing all of the contributions.

It should also be noted that when multiple incident outcomes are possible, it might be more conservative (and reduce the effort required) to use the expected maximum consequence. For instance, in example 8.3, the maximum expected consequence would be a vapor cloud explosion. A conservative consequence estimate of PLL=13.7 could have been made without going through the effort of performing the consequence modeling required for the flash fire.

Example 8.3

Problem: A release of propane can result when the packing gland of a compressor is damaged. Based on experience in this particular application, it has been determined that there is a 30 percent probability the released material will ignite. If the material does ignite, there is only a 5 percent chance an explosion will occur. Quantitative consequence analysis has determined that the following incident outcomes are possible and will have the associated consequences shown:

Vapor cloud explosion	PLL=13.7
Flash fire	PLL=8.4

What is the average consequence of a propane release from a damaged compressor packing gland?

Solution: The following graphic shows the event tree describing the potential incident outcomes. The initiating event is the release of propane, and the branches that determine the potential outcomes are probability of ignition and probability of explosion. The potential outcomes are vapor cloud explosion, flash fire, and no significant consequence. Calculate the probability of each outcome by using probability multiplication.

Once the probabilities are determined, calculate the contribution to the average from each incident outcome by multiplying each consequence by its probability. In this case, the consequence is given in terms of probable loss of life (PLL). The overall average consequence is then determined by summing the contributions from each incident outcome.

INIT EVENT	BRANCH 1	BRANCH 2	OUTCOME	PLL	CONTRIB.
Propane Release	Ignition	Explosion			
		5%	0.015	13.7	0.2
	30%	TRUE	VCE		
	TRUE				
1		95%	0.285	8.4	2.4
		FALSE	Flash Fire		
	70%		0.7	0	0.0
	FALSE		Nothing		
		Average Consequence in terms of PLL:			2.6

8.7 Summary

This chapter described the method of event tree analysis that is commonly used to analyze process accidents. Event tree analysis is the most straightforward of the fault propagation modeling techniques and is very powerful in situations where the series of events do not include complex redundancy or on-line repair schemes.

These trees are analyzed from left to right, starting with the initiating events and working to one of several outcomes through a series of branches. Initiating events by definition start the chain of events that leads to the potential unwanted accident. Some examples of initiating events include the failure of process components and improper operation of the process. Branches are the sets of events that determine which different outcomes are possible. Although many branch event sets are complementary, this is not always the case. Branch events are always described in terms of probability. Once you have generated all of the branches, list the resulting potential outcomes on the right-hand side of the diagram. The probability or frequency of an outcome is a logical combination of the initiating event and the branch events on the path that connects the initiating event to that outcome.

Outcome probabilities are calculated by probability multiplication because the outcomes are a logical "AND" of the initiating event and the branch events leading to the outcome. In other words, the frequency of a particular outcome is the frequency of the initiating event multiplied by the probabilities of the branch events leading up to it.

In addition to their effectiveness in calculating outcome frequencies, event trees are also widely used to calculate average incident outcomes. When a number of incident outcomes are possible from a single event, you can determine the probability of each outcome by using the event tree and then multiplying the consequence of that outcome to determine its weighted contribution. The sum of the weighted incident outcome contributions is then the average consequence.

8.8 Exercises

8.1 Draw an event tree and quantify the outcomes for overfilling a flammable materials tank. Define the initiating event as the delivery of the flammable material. Events that will determine the potential outcomes include:

1. The tank may not have enough room to hold the delivery (10%).
2. The operator may not detect that not enough room exists before starting the transfer (5%).
3. The operator may not supervise the transfer and thus will not detect the high level in the tank (15%).

8.2 A process incident may result in the breakage of a pipe elbow, which will then allow a mixture of ethane and propane to escape continuously at high velocity. If the release is ignited immediately (5% chance), the incident will result in a jet fire causing $5 million worth of equipment damage. If the ignition is delayed, the release will result in either a flash fire or vapor cloud explosion. Delayed ignition has a probability of 25%. A flash fire would result in $850,000 of equipment damage, while a vapor cloud explosion would result in $45 million in equipment damage. The calculations from a very detailed computational fluid dynamics model of the process area and release have yielded an estimate that 15% of the delayed ignition events will result in a vapor cloud explosion rather than a flash fire. What is the average consequence of this pipe elbow breakage incident?

8.9 References

1. Lees, F. P. *Loss Prevention for the Process Industries.* London: Butterworth and Heinemann, 1992.

CHAPTER 9

Layer of Protection Analysis

Layer of protection analysis (LOPA) is a special form of event tree analysis that is optimized for the purpose of determining the frequency of an unwanted event, which can be prevented by one or more protection layers. By comparing the resulting frequency to the tolerable risk frequency from chapter 3, we can finally select the appropriate safety integrity level (SIL) (the subject of chapter 10). LOPA uses initiating events in much the same way as event tree analysis, but it requires that they be expressed in terms of frequency. The protection layers in LOPA are analogous to the branches in event tree analysis. In LOPA, each branch is always a set of complementary events in which the protection layer either operates successfully or fails. We calculate the probability of unwanted events much as we do in event tree analysis, with the exception that we are interested in only one outcome. Thus, only the one harmful outcome frequency is usually ever calculated.

When conducting LOPA, it is sometimes necessary to consider multiple initiating events. This is accomplished by analyzing each individual event with a separate LOPA diagram and then selecting the SIL based on the event that has the most stringent requirements.

9.1 LOPA Overview

LOPA is a variation of event tree analysis that is limited and optimized for a specific situation. The specific situation involves an initiating event that can lead to an unwanted incident, but one or more protection layers can also prevent the incident from occurring. Just as in the more general form of event tree analysis, the initiating event starts the chain of events that leads to the unwanted impact (see figure 9.1). However, in LOPA the initiating events are always described in terms of frequency. The different layers of protection are also considered much as the branches in an event tree. When the analyst encounters a protection layer branch and the result of the protection layer is success, then that branch of the analysis is terminated by the successful no-incident outcome. Pursuing this to its logical conclusion, only two outcomes are then possible: the harmful incident or no event. Of those two outcomes, only the harmful incident frequency is calculated since it is the only outcome of any real interest.

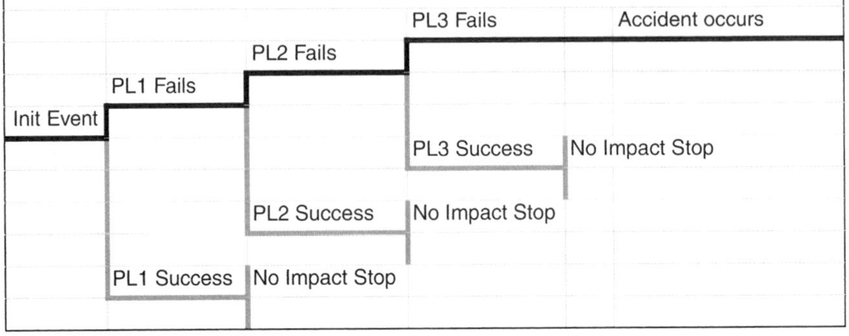

Figure 9.1 Diagram of Typical Layer of Protection Analysis

PL3 Fails
Accident occurs
PL2 Fails
PL1 Fails
Init Event
PL3 Success
No Impact Stop
PL2 Success
No Impact Stop
PL1 Success
No Impact Stop

Figure 9.1 is a typical diagram illustrating a layer of protection analysis. As with the diagrams used for event tree analysis, the LOPA diagram is viewed from left to right and begins with the initiating event. In this case, there are three layers of protection subsequent to the initiating event. Each layer of protection is a set of complementary events: either protection-layer failure or protection-layer success. If the protection layer succeeds, there is no impact and the analysis of that success branch ends. If the protection layer fails, the analysis of the failure branch proceeds to the next protection layer or to the relevant harmful outcome.

9.2 Protection Layers and Mitigating Events

Each protection layer in a LOPA analysis is analogous to an event tree branch, except that the former are more restricted. While event tree branches can contain any type of independent event set, LOPA protection layers must be complementary events in which, as we have said, the protection layer either fails or operates successfully. The probability associated with the failure path is the probability of failure on demand (PFD) of the protection layer, or the probability of a mitigating event.

In this discussion of LOPA we will use the term *protection layer* somewhat loosely to include all events that decrease the probability that an incident will occur. So in addition to the engineered systems specifically designed to reduce risk, a protection layer will also include an incidental event that happens to reduce risk but may not have been specifically designed to do so. Many risk analysts separate these two categories of risk mitigation into *protection layers* and *mitigating events*. Protection layers are the specifically engineered devices, while mitigating events are things that reduce risk but are not specifically designed to do so, such as the probability of ignition and the probability of explosion. Making this additional classification is up to the user, but the most important consideration is to establish a clear way to include each and every one of these independent protection layers and mitigating events exactly once in the analysis. Whether the protection comes from an engineered protection layer or a mitigating event, the risk reduction is handled the same way in the LOPA analysis.

Formally defined protection layers that are engineered to reduce risk should meet the following criteria.

- ***Specificity*** – An independent protection layer must be specifically designed to be capable of preventing the consequences under consideration.
- ***Independence*** – The protection layer must operate completely independently of all other protection layers; no common equipment can be shared with other protection layers.
- ***Dependability*** – The device must be able to dependably prevent the consequence from occurring. Both systematic and random faults need to be considered when the device is designed.
- ***Auditability*** – The device should be proof tested and maintained. These audits of operation are necessary to ensure that the specified level of risk reduction is being achieved.

9.3 LOPA Quantification

The analyst quantifies the outcomes of the layer of protection analysis using the same process as in event tree analysis. Again, in this case, only one outcome is of interest: the occurrence of the unwanted accident. The frequency at which a potential accident is initiated but which a layer of protection prevents from occurring is typically of no interest to the analyst and is usually not calculated.

The probability of the unwanted accident is logically related to the initiating event and the protection layers by a logical AND. The resulting accident frequency is the frequency at which the initiating event occurs and all of the protection layers fail to prevent the accident. Since the relationship is a logical AND, we use probability multiplication to calculate the outcome frequency. The outcome is the initiating event frequency multiplied by the probability of failure on demand (PFD) of all of the protection layers. In cases where mitigating events are also considered, the probability of each mitigating event failing to prevent the accident is also considered through probability multiplication in the same way as for the protection layers.

Example 9.1

Problem: A process hazards analysis (PHA) has determined that a distillation column can rupture, causing the release of flammable materials and a pool fire. The PHA identified that the initiating event for the column rupture is loss of cooling water, which has been determined to occur with a frequency of 0.5 per year. The layers of protection that can prevent the cooling water loss from propagating into a column rupture include the following:

Process Design	PFD=0.01
Operator Response	PFD=0.2
Pressure Relief Valve	PFD=0.07
Ignition Probability	PFD=0.3

Solution: The initiating event, loss of cooling water, starts the diagram on the left. We then list the protection layers after the initiating event, with the failure branch events leading to either the next protection layer or the final outcome, and the success branch events leading to the no-event outcome. The unwanted accident outcome, or the column rupture and fire, is the outcome of interest on the right side of the LOPA diagram. The other outcome is "no event."

We include the probabilities for the initiating event frequency and protection layer failure in the LOPA diagram, as in the following table. The no-event outcome is of no interest, so we do not calculate its frequency. The rupture and fire is the outcome that we calculate. We use probability multiplication to calculate this outcome frequency as follows:

$$\text{Frequency} = 0.5 \times 0.01 \times 0.2 \times 0.07 \times 0.3 = 2.1 \times 10^{-5}$$

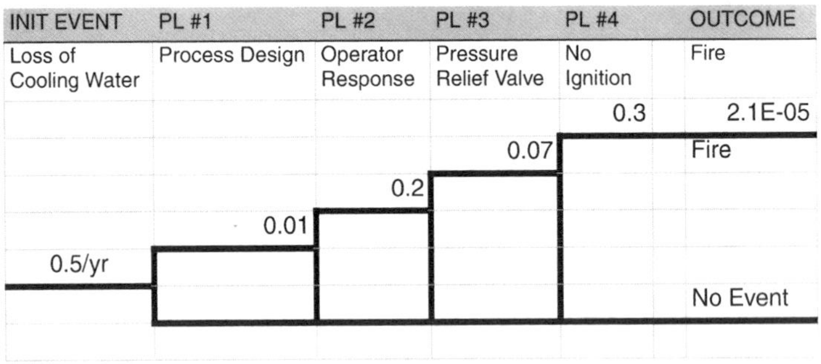

INIT EVENT	PL #1	PL #2	PL #3	PL #4	OUTCOME
Loss of Cooling Water	Process Design	Operator Response	Pressure Relief Valve	No Ignition	Fire
				0.3	2.1E-05
			0.07		Fire
		0.2			
	0.01				
0.5/yr					
					No Event

9.4 Typical Protection Layers

While no two situations are identical, a few protection layers and mitigating events should always be considered when performing a layer of protection analysis in the process industries:

- Basic process control system (BPCS)
- Operator intervention

- Use factor
- Mechanical integrity of vessel
- Physical relief device
- External risk reduction facilities
- Ignition probability
- Explosion probability
- Occupancy

One vital aspect of layer of protection analysis is ensuring that each layer of protection is considered once and only once, as required by the independence criterion described in section 9.2. Layers of protection must be considered as functioning on a completely independent basis because we use probability multiplication in the analysis. For instance, when considering the mechanical integrity of the vessel, you must work with the situation in which the relief device does not prevent the accident when determining the probability of failure upon demand (PFD) for the mechanical integrity. Thus, you must consider the probability that either the mechanical integrity layer will fail because of shock below the threshold of the relief valve or that the vessel mechanical integrity layer will fail when the relief valve does not function properly to relieve a high-pressure condition. Otherwise, the probability of failure on demand will be much too low and will give a dangerously low prediction of the overall accident frequency. In general, this means that you must focus on the way each layer of protection is sequenced in the LOPA. You must be sure to evaluate every layer under the conditions that all of the preceding layers have failed to stop the initiating event from propagating. Incorporating this discipline in the analysis will help prevent you from generating dangerously low predictions of overall failure rates, which can lead to potentially inadequate safety system designs.

9.4.1 Basic Process Control System

The basic process control system, or BPCS, is considered to be a layer of protection under certain circumstances. A BPCS can often be configured to mimic the actions of a safety instrumented system (SIS) and to perform those actions before the SIS takes action. The following are the three criteria most commonly used to determine whether a BPCS can be considered as a protection layer:

- The BPCS and SIS are physically separate devices, including sensors, logic solver, and final elements.
- Failure of the BPCS is not responsible for initiating the unwanted accident.
- The BPCS has the proper sensors and actuators available to perform a function similar to the one performed by the SIS.

When a BPCS is used as a protection layer, its performance is measured by its probability of failure on demand (PFD). The PFD of an instrumented system is determined by reliability analysis. As an option, many protection layer databases contain data on failure rates for entire control loops.

It is also important to be sure to clarify the BPCS action with respect to operator intervention. Note that the BPCS layer of protection is generally associated with an automatic action that does not require any direct operator intervention. If there is an alarm or other indication to prompt the operator to act and that action is what provides the protection, then it is generally best to consider that fact in the operator intervention LOP. The most important thing is to be sure not to take double credit for any one layer of protection.

In any case, the amount of credit for BPCS effectiveness is limited by the lower boundary of an SIL 1 safety instrumented function (SIF). When performing layer of protection analysis, ensure that the PFD for a BPCS is not less than 0.1. If you desire a PFD of less than 0.1 for a BPCS, then you should design, install, and maintain the BPCS as an SIS according to IEC 61511 and/or ANSI/ISA-84.01-1996. This will effectively make the BPCS an SIS.

9.4.2 Operator Intervention

A trained and attentive operator is the first line of defense against accidents. Process operators are responsible for monitoring the state of the process and making changes to process set points that will not only keep the process in a safe state, but also maintain product quality and maximize profitability. To perform this safety task, the operator must be well trained. He or she must know the normal operating ranges of all of the process variables as well as how set point changes will impact the different variables. To accomplish all of these complex tasks effectively, the operator must also be well rested and motivated.

Operators help to keep the process in a safe state in two main ways. First, an effective operator will ensure that the process stays in control. If the process starts to veer toward an out-of-control situation, an attentive operator will adjust process set points to keep all operating variables in their normal operating range. Second, an operator is also capable of stopping the process or taking other similarly severe actions to move the process to a safe nonoperating state when process control is lost.

In order for an operator to shut down a process when it goes out of control, several criteria must be satisfied. The operator first must be alerted that process variables have gone out of their normal operating range either by automatic alarms or by the operator's systematic review and recording of all key process variables. When the process variables are out of their normal operating range, the operator then must be able to diagnose the root cause of the problem. Out-of-range variables are a symptom that may have a complex relationship to the actual root cause. The task of diagnosis is facilitated by training in which the operator understands the process and performs practice drills on the proper response to alarms or sets of alarms. Finally, the operator must also take an action to move the process to a safe state. This action depends not only on the operator's diagnosis, but also on the proper functioning of the mechanical equipment that shuts the process down and moves it to a safe state.

Human factors have a great deal of impact on an operator's ability to perform a shutdown. An operator's familiarity with the task being performed is critical. The most effective operators will have been drilled on the series of actions they must take in the case of certain critical alarms. This type of training makes reaction automatic, so no complex diagnosis or detailed problem-solving is required. An operator who has been drilled to respond to a certain alarm with specified actions provides the most effective type of human response.

The sheer complexity of the task of moving the process to a safe state also impacts operator effectiveness. While an operator can perform a single push-button response to an alarm very effectively, a long series of actions that requires process feedback in between steps and complex decisions along the way will often present problems. For instance, depressuring a Hydrocracker is an example of a high-complexity shutdown. To bring a Hydrocracker reactor to a safe state after a runaway reaction is detected, the operator would have to stop the recycle gas flow by stopping the recycle gas compressor, stop the hydrocarbon feed flow by stopping the feed pump, and open a depressuring valve to vent the reactor. These initial steps to bring the process to a safe state must then be followed by additional actions to make sure that new hazards are not introduced, such as closing hydrocarbon feed valves to prevent backflow of the reactor contents into the low-pressure feed system.

The amount of time the operator has available for diagnosis will also influence his or her ability to take the proper action. Unless the operator is drilled to automatically take a specific action upon receiving a specific alarm or set of alarms, he or she will have to make more decisions. These decisions require the time to assess the situation, determine the root cause, identify the proper course of action, and then perform that action. Some experts estimate that this type of involved decision-making will take an operator about ten minutes from the time the alarms are received to the point when the operator can act to prevent an accident.

Critical, of course, to the operator's ability to respond is receiving an indication of out-of-control conditions. Variables that have an impact on safety should be equipped with alarms. If a process variable is not alarmed, the operator will have little chance of detecting the out-of-control situation and responding in time.

The operator's workload will also determine how quickly he or she can take preventive action. This stems both from the reduced likelihood that the operator will be monitoring the area where the problem starts and from the greater difficulty they will have promptly deciding how to prevent the accident.

The physical condition of the operator, including age and amount of rest, as well as the ergonomics of the workplace will also have a large impact on safety. These factors control alertness, which determines the amount of time the operator has to respond. Poor workforce morale also tends to hurt overall effectiveness.

Several methods have been devised to quantitatively estimate an operator's efficacy. This "operator PFD" is expressed in terms of the probability that the operator will fail to respond correctly to a given set of conditions. Most of these techniques consider the type of task that the operator is being asked to perform and the error-producing conditions that might prevent the operator from responding correctly. In his *Loss Prevention for the Process Industries,* Lees surveys a range of techniques, giving the strengths and limitations of each. The Human Error Analysis and Reduction Technique (HEART) is particularly well suited to estimating an operator's ability to properly respond to a critical situation.

Table 9.1 Simplified Technique for Estimating Operator Response

Category	Description	PFD
1	**Normal Operator Response** - In order for an operator to respond normally to a dangerous situation, the following criteria should be true: • Ample indications exist that there is a condition requiring a shutdown • Operator has been trained in proper response • Operator has ample time (> 20 minutes) to perform the shutdown • Operator is ALWAYS monitoring the process (relieved for breaks)	0.1
2	**Drilled Response** - All of the conditions for a normal operator intervention are satisfied, and a "drilled response" program is in place at the facility. Drilled response exists when written procedures, which are strictly followed, are drilled or repeatedly trained by the operations staff. The drilled set of shutdowns forms a small fraction of all alarms where response is so highly practiced that its implementation is automatic. This condition is rarely achieved in most process plants.	0.01
3	**Response Unlikely** - All of the conditions for a normal operator intervention probability have not been satisfied.	1.0

A simplified technique for estimating operator response, such as the one outlined in table 9.1, is the one commonly used in the process industries as part of the SIL selection process. This method categorizes the human factors surrounding the potential accident scenario and assigns a nominal operator unreliability to the situation. The operator effectiveness is typically quoted in terms of probability of failure, which can be used directly in the LOPA analysis as a layer of protection.

Example 9.2

Problem: A HAZOP study of a process plant has identified that an explosion could occur because of a runaway reaction in a batch reactor. In this case, the reaction runaway develops over a period of 30 minutes. When a runaway reaction occurs, the process temperature will exceed the high-temperature alarm set point for a full 20 minutes before the explosion is expected to result. The operators have been trained in the proper response to this alarm. Using the table for estimating simplified operator response (table 9.1), estimate the operator's ability to prevent the explosion in the two cases that follow.

Case #1 – A plant has only one operator typically monitoring the process from the control room. But once per shift, he takes measurements from local gauges in the field, which requires about 60 minutes of effort.

Case #2 – There are two operators in the control room. While each of the operators is required to take outside measurements, one of the operators is continuously monitoring the control panel.

Solution: Based on the problem statement, several conditions for normal operator response are met in both cases: (1) there is ample indication that a condition requiring operator action exists, (2) the operator has been trained in the proper response, and (3) the operator has ample time to perform the shutdown. In neither case could the response to the alarm be considered a drilled response based on the information provided. Both are simply normal responses to any one of a large number of alarm conditions, and are therefore given no special status.

In case #1, the operator often leaves the control panel in the control room unattended for long periods of time (up to 60 minutes) when taking field measurements. As such, case #1 does not meet the normal operator response condition, in which the operator continuously monitors the process. Since case #1 does not meet normal operator response or drilled response, the category selected for this scenario is "response unlikely" with a probability of failure on demand of PFD=1.0.

In case #2, the operator is always relieved in his control panel monitoring tasks, so the control panel is continuously attended. As such, case #2 meets all of the conditions required for normal operator response. Since case #2 meets the criteria for normal operator response but not drilled response, the category selected for this scenario is normal operator response with a probability of failure on demand of PFD=0.1.

The simplified technique, as presented in table 9.1, separates operator effectiveness into three categories, which can be roughly described as (1) normal operator response, (2) drilled response to predetermined alarm conditions, and (3) operator ineffectiveness.

A typical operator response is assigned a probability of failure of 0.1, which is equivalent to an SIL 1 safety instrumented system. The following constraints must be met for this effectiveness to be valid: (1) the operator has ample indications that a situation requiring a shutdown has occurred; this warning should be in the form of alarms; (2) the operator has been trained in the proper response to this specific scenario; (3) the operator has ample time to assess the situation and perform the shutdown, usually more than twenty minutes; (4) an operator is *always* monitoring the process. This last rigorous requirement implies that the staffing procedure allows the main operator to be relieved by an alternate during all breaks. If all of these conditions are met, then the operator response is considered typical and assigned a PFD of 0.1.

A drilled response is assigned a probability of failure of 0.01, which is equivalent to an SIL 2 risk reduction factor. This scenario is assigned an effectiveness of 0.01 because of the highly trained and nearly automatic nature of the operator response. For this situation to exist, all of the conditions for a typical response must be met first. In addition, the response must be included in written operating procedures and executed by highly trained and motivated operators. This is a very stringent condition that is rarely met in most process plants. This type of response can only be performed for a few critical loops since the operator will not be able to provide this type of response for every loop in the plant.

If neither of these two categories is achieved, then the operator is considered to be ineffective at performing a manual shutdown under safety-critical conditions, and the PFD is assigned a value of 1.0.

9.4.3 Use Factor

The use factor defines the fraction of time in which the hazardous process is in operation or the hazard is present in the system. The use factor is an important consideration when estimating the frequency of an accident when the hazard only exists when the process is in operation or even when it is in a certain mode of operation. For example, consider a boiler that is only operational six months of the year. The hazard is only present half of the time that the same hazard would be present for a boiler that is operated continuously. It therefore follows that the risk posed by the boiler that operates for half the year should be half of the risk posed by the continuously operated boiler.

We determine the use factor by dividing the time that the process is in the hazardous mode of operation by the total time. For processes operated seasonally, one calendar year (i.e., 8,760 hours) is usually the best total time base. On the other hand, some batch operations require 12 hours to complete, from charging the reactants to removing the final products. For them, 12 hours will be the time base if the batches are continuously run one after another. If the batches are not continuously run, the average amount of downtime in between batches must also be considered as part of the time base.

Example 9.3

Problem: A dedicated fuel gas-fired boiler creates steam to heat the engineering offices of a refinery. The offices are located in an area with a mild climate, so steam heating is only required an average of 14 weeks every year. What is the use factor of the boiler process?

Solution: Based on this situation, we calculate the use factor as follows.

First, determine the duration of time of one full cycle. Since this example is a seasonal operation, the cycle time is assumed to be one year. The second step is determining the amount of time that the hazard is present. The problem states that the hazard is present for 14 weeks out of an average year. We then calculate the use factor by dividing the time the hazard is present by the total cycle time.

Use Factor = Presence Time / Cycle Time = 14 weeks / 52 weeks = 0.27

9.4.4 Mechanical Integrity of the Vessel

The mechanical integrity of the vessel is an often overlooked but highly effective protection layer capable of guarding against many overpressure and over-temperature events. During process hazards analyses (PHAs), engineers often specify safety functions to protect against hazards that the process equipment is already designed to withstand. For instance, an initiating event might be the rupture of tubes in a shell and tube steam reboiler of a distillation column. In this scenario, the distillation column was designed to withstand full steam header pressure, but the introduction of high-pressure steam will be a shock to the system nonetheless.

One could argue that since the process equipment is designed to withstand this condition, the overpressure or over-temperature scenarios are not credible events. This is a valid viewpoint, but many organizations take a more conservative stance. Even though the process is designed to withstand a certain temperature or pressure, a sudden, dramatic increase in temperature or pressure, even below the design limits, could reveal latent faults or defects and damage equipment.

If an organization believes that the rupture of a piece of equipment because of shocks below its design conditions is a credible accident scenario, then it must consider the possibility that the vessel design is sufficient to withstand the shock as a protection layer. It is difficult to estimate the effectiveness of this protection layer because the data to quantify this scenario is lacking. A rule of thumb for quantifying the failure probability of this protection layer is to assign the normal failure probability for an entire year to this single event using equation 5.11. In this way, the probability of vessel failure because of the single stress incident is the same as the probability of the vessel failing during one year of normal operation.

Example 9.4

Problem: An SIS is being considered to prevent the rupture of a distillation column. The initiating event is the rupture of the steam reboiler tubes, which will suddenly raise the pressure in the vessel from the normal operating pressure of 150 psig to 250 psig. The vessel is designed to withstand 300 psig at the temperature of the steam. Estimate the probability that the vessel will not be able to withstand this sudden high-pressure and high-temperature shock.

Data from OREDA-92 indicate that the failure rate of a pressure vessel with the failure mode of "critical – significant leakage" is 0.35 failures per million hours.

Solution: Although the vessel is designed to withstand the temperature and pressure indicated for the hazardous situation, there is a possibility that the sudden shock might reveal some latent faults. We estimate the failure probability by using the rule of thumb that a sudden shock will have the same probability of causing a failure as one year of normal service.

Using equation 5.11 with a failure rate of 0.35 failures per million hours and a mission time of one year yields the following:

$$PFD = 1 - e^{-0.35 \times 10^{-6} \times 8760} = 3.1 \times 10^{-3}$$

9.4.5 Physical Relief Devices

Physical relief devices are items of non-instrumented mechanical equipment that perform an action to relieve pressure when the normal operating range of temperature or pressure has been exceeded. Physical relief devices include pressure relief valves, thermal relief valves, rupture disks, rupture pins, and high-temperature fusible plugs.

Relief valves open to let process material escape relatively harmlessly to secondary containment or to the atmosphere when excess pressure occurs in the process. There are a wide variety of relief valve styles and technologies, but most are either direct-acting or pilot-operated. Direct-acting relief valves are opened by the force of the process fluid acting against a spring in the valve. When the force created by the process exceeds the force of the spring, the valve opens, relieving process pressure. Once the force from the pressure drops back below the relieving point, the valve returns to the closed position. Pilot-operated relief valves operate much like direct-acting valves, but they use the process fluid, via a pilot mechanism, to help open and close the valve.

Rupture disks are engineered failure points in the overall integrity of a vessel. Rupture disks are designed and selected so that the disk will burst at a lower pressure than the design pressure of the vessel that is being protected. This allows the contents of the tank to be safely relieved to the atmosphere or secondary containment. If they are selected and installed properly, rupture disks have an extremely low probability of dangerous failure, which is essentially limited to plugging the path to the disk or high backpressure. While rupture disks are extremely effective, their application is often limited because of nuisance trips and the large process downtimes often required to replace a blown disk.

Rupture pins are also engineered failure points and have been gaining acceptance recently. Here, a deformable pin bends against an activating force that is created by the process pressure to open a path to a vent or secondary containment for the process fluid. Because their operating concept is similar to disks, they also have a low dangerous failure rate. Moreover, nuisance trips and process downtime instances are not as severe with rupture pins as with traditional rupture disks, although this depends on the particular installation.

Fusible plugs also operate much like rupture disks, but they are activated by high temperatures instead of high pressures. Fusible plugs are often used in transportation containers where the only expected mode of overpressure is external fire, such as in chlorine cylinders. A fusible plug is a process connection that contains a solder plug whose melting point is selected so it will melt and allow the contents of the vessel to escape before an external fire causes an explosive overpressure event.

The probability that a physical relief device will fail varies with the type of device. The failure rates for the various designs of relief valves can be found in numerous databases and reference books. The failure probability of a rupture disk is largely a function of the likelihood of the path to the disk becoming plugged and/or of excessive backpressure. Human factors surrounding the installation of a rupture disk may also play a role in its failure probability, depending on the type used. Failure rates for fusible plugs are also very low, on the same order as that of rupture disks. You should determine the failure rates for fusible plugs by consulting the device's manufacturer.

It is important to note that the physical relief devices described here all depend on the relief device and the downstream flow path being large enough to relieve the hazardous condition. Care must be taken during the design of these devices and their downstream piping to ensure that they can reduce the pressure faster than the highest potential buildup rate. Otherwise, the pressure in a system can continue to build even though the relief valve is open. This would cause the presumed protection layer to fail just as surely as if it had never opened in the first place.

9.4.6 External Risk Reduction Facilities

External risk reduction facilities (ERRF) are "active" protection layers that perform their actions after the chemical has already been released into the atmosphere. They are active systems in that some form of energy must be applied to cause the ERRF to operate. The opposite of active systems are passive systems, such as dikes and containment vessels, which are assumed to never technically fail but only function to lessen the severity of a release consequence.

For an ERRF to be considered a layer of protection, it must completely prevent the consequence in question. If the ERRF only lessens the severity of the consequence, it cannot be considered here. In that case, the risk reduction effect of the ERRF should be considered as part of the consequence analysis. An example of an ERRF is a water curtain that is activated when chlorine is released. If this water curtain is designed to completely prevent the released chlorine from exiting the curtain and impacting personnel, then we can consider it in the layer of protection analysis.

An ERRF's probability of failure is determined by analyzing its components. In some cases, this will require a fault propagation model such as a fault tree. For instance, if a water cannon system is used for an ERRF, its failure may be a function of the failures of a series of valves and pumps.

As with the physical relief devices we described earlier, ERRF systems must be designed and sized with extreme care. One specific instance when this was not done was for a water quench system in Bhopal, India, in 1984. The Methyl Isocyanate (MIC) water quench system was severely undersized for the release that took place. In fact, the release was so much larger than the water quench design that the quench reaction energy release actually exacerbated the disaster by dispersing the remaining unreacted MIC more widely.

9.4.7 Ignition Probability

In many cases when a flammable material is released into the atmosphere, the outcome effect of explosion or fire is never realized because the released material is never ignited. The probability that a release will ignite depends on a number of factors, including the chemical's reactivity, volatility, auto-ignition temperature, and physical state as well as the potential sources of ignition that are present.

Ignition occurs when the local energy of the fuel and oxidant mixture is raised above the level where the combustion reaction produces more heat than the combustible mixture loses. This causes the combustion reaction to become self-sustaining. This local energy often takes the form of temperature, and the critical value is called the *auto-ignition temperature*. However, other forms of energy such as shock or an electrical arc can also cause ignition. The chemical's reactivity impacts the amount of energy it takes to begin combustion. More reactive chemicals such as ethylene oxide are more likely to ignite than low-reactivity chemicals such as propane. Combustion is usually a vapor phase reaction; thus, in order for liquids to ignite and burn the liquid must volatize. For this reason, materials with a high vapor pressure, such as pentane, are much more likely to ignite than materials such as fuel oil or diesel fuel, which have lower vapor pressure. For the same reason, flammable gases are much more likely to ignite than flammable liquids. When a material is released above its auto-ignition temperature, it will ignite immediately upon contact with oxygen.

Unless the auto-ignition temperature has been surpassed, it will take another source of energy for ignition to start the combustion. As one would expect, more sources of ignition will result in a higher probability of ignition. In a process environment, sources of ignition include fired heaters, flares, electric motors (e.g., pump motors), hot work, static electricity, and other equipment that might generate a spark or have a hot surface. While good engineering practices will limit the probability that ignition sources are present, they cannot be entirely removed. *Loss Prevention for the Process Industries* contains a lengthy discussion of ignition probabilities.

9.4.8 Explosion Probability

Explosion probability indicates the probability that a release of flammable material will explode, given that it has already ignited. Not all ignitions of flammable vapor clouds will result in an explosion. For a blast to result from vapor cloud combustion, a reasonable amount of obstructions and confinement must exist to cause the flame front to burn turbulently and reach sonic velocity. In many cases, the vapor cloud will burn in a laminar fashion and not develop any significant overpressure that would then cause a blast wave. This laminar burning vapor cloud is called a *flash fire*, which may have significant consequences of its own. We discussed flash fires and vapor cloud explosions in more detail in section 6.4.

In most cases, explosion probability should be ignored in a LOPA. The only exception is where the flash fire will be completely contained inside an unoccupied enclosure and no receptors will be impacted. Otherwise, the absence of an explosion does not prevent a consequence; it simply decreases the consequence compared to an explosion. In the case where a flash fire is a less severe consequence than an explosion, you should either assume that all ignitions will lead to explosions, which is a conservative assumption, or calculate an "average" consequence using an event tree as described in section 8.6.

9.4.9 Occupancy

Occupancy is a measure of the probability that the effect zone of an accident will impact one or more personnel. We determine this probability by using plant-specific staffing philosophy and practice. It is important to note that many quantitative consequence analyses, such as those created by consequence modeling software applications, already incorporate the probability that the hazard zone is occupied. As such, you should leave the occupancy probability as 1.0 to avoid "double accounting" for occupancy when using quantitative consequence analysis to select SILs.

We recommend considering occupancy as part of the consequence analysis rather than using occupancy probability in the LOPA. The limitation of occupancy probability is that by definition it considers a single fatality. If the consequence could cause multiple fatalities, the occupancy probability method could lead to erroneous results.

Example 9.5

Problem: A batch process requires that an operator manually add reactants to a reactor. The entire process is enclosed within a structure that will contain any potential release of toxic reaction byproducts. An SIF is being installed that stops the reaction before toxic materials are released to the enclosure so as to protect the operator. The potential for the release of toxic material is continuously present, but the operator typically only spends 30 minutes in the enclosure during an 8-hour shift. What occupancy should be used for this situation in a layer of protection analysis?

Solution: The occupancy is the probability that a receptor will be in the effect zone to sustain the effects of a release. In this case, the effect zone would be the entire enclosure. The operator's occupancy of the enclosure is 30 minutes of an 8-hour shift, or:

$$P_{occcupancy} = 0.5/8 = 0.06$$

Note that this solution assumes that there is an equal probability that the release will occur whether the operator is present or not. This occupancy must be adjusted upward if the release is more likely to occur during the part of the process when the operator is inside the enclosure.

9.5 Multiple Initiating Events

The LOPA procedure we have described up to this point assumes that only one initiating event causes the accident. However, accidents often have multiple potential triggers that can propagate to an unwanted accident. Consider a fired heater that uses natural gas as a source of fuel. The loss of the heater flame without isolating the heater's fuel gas supply could eventually cause the consequences of vapor cloud explosion or flash fire. The initiating events that can cause this scenario include, for example: (1) a momentary drop in fuel gas pressure, (2) a momentary high-pressure spike, (3) a pocket of inert gas in the fuel line that extinguishes the flame, (4) a ratio of fuel gas to air at the burner that is either too high or too low to support combustion. A number of other initiating events may also result in this scenario depending on the specific heater configuration and fuel source.

Not only are multiple initiating events possible, but the protection layers are not always effective against all initiating events. A typical fired-heater safety instrumented function (SIF) might use a low-pressure switch and a flame scanner as inputs to shut off the fuel gas supply. The flame scanner protection layer will be able to detect the existence of all four of the initiating event failure modes we mentioned in the last paragraph. However, the low-pressure switch protection layer will only be effective against the momentary pressure dip, and possibly the high air-to-fuel-pressure ratio triggers.

Despite these complexities, LOPA can be used very effectively in situations where multiple initiating events exist. When this is the case for a single hazard, break the hazard into parts and consider each initiating event as a separate LOPA. Once you have performed each of the analyses, you should have a frequency of incident based on each individual initiating event. You then determine the overall frequency of the incident produced by all initiating events by summing the frequencies caused by each of the individual initiating events.

Example 9.6

Problem: An SIS is being designed to prevent the rupture of a distillation column because of excess pressure. Excess pressure in the vessel can occur as the result of two initiating events: (a) rupture of a steam tube in the column's reboiler releasing live steam directly into the process, and (b) failure of the cooling water pump that supplies cooling water to the column's condenser. After LOPA was performed on both of these situations, it was determined that the following accident rates would result from each initiating event:

Steam tube initiated column rupture: 1.5×10^{-3} per year

Cooling water pump failure initiated column rupture: 9.6×10^{-3} per year

What is the overall rate at which ruptures can be expected to occur as a result of all initiating events in this scenario?

Solution: The overall incident rate resulting from multiple initiating events is the sum of the rates contributed by each individual initiating event:

$$\text{Incident Rate} = 1.5 \times 10^{-3} + 9.6 \times 10^{-3} = 1.11 \times 10^{-2} \text{ per year}$$

9.6 Summary

This section explored layer of protection analysis (LOPA), a type of fault propagation modeling that is derived from event tree analysis methods. LOPA is specifically optimized for determining the frequency of process accidents that result from initiating events that can be prevented by one or more protection layers.

When performing a LOPA, the relevant initiating events are identical to those that would be used in event tree analysis, except that they are always represented in terms of frequency. LOPA protection layers are similar to event tree branches, but more restricted. A protection layer branch is always a complementary set of events in which one leg—the failure of the protection layer—leads to the unwanted accident while the other leg leads to the no-accident outcome.

LOPA diagrams are quantified in the identical way as event trees, but only one outcome, the unwanted accident, is usually calculated. The method is based on the assumption that each layer of protection acts independently of the others, so the calculation is performed using probability multiplication. The accident frequency is the initiating event frequency multiplied by the probability of failure on demand of all of the protection layers.

A number of protection layers and mitigating events are commonly employed in the process industries. These include the basic process control system, operator response to alarms, use factor, mechanical integrity of process vessels, physical relief devices, external risk reduction facilities, ignition probability, explosion probability, and occupancy. You should consider all of these protection layers each time you perform a LOPA, but you should take extreme care to ensure that credit for a protection layer is not counted twice.

LOPA analysis can be extended to include the numerous industrial accidents caused by multiple initiating events. When the outcomes are the same, we add the resulting outcome frequencies from each of the multiple initiating events together to give the total frequency of that outcome.

9.7 Exercises

9.1 What type of probability math is used to calculate the outcome frequency of a LOPA diagram?

9.2 What are the differences between LOPA and traditional event tree analysis?

9.3 Explain the difference between a mitigating event and a formal protection layer. Give an example of each.

9.4 What are the four criteria necessary for a formally defined protection layer?

9.5 Name six typical protection layers or mitigating events that are commonly applied in the process industries.

9.6 An SIF is being designed to prevent a pressure and temperature shock that is within the temperature and pressure limits of a process vessel. Operational data shows that the expected failure rate (rupture) of this type of vessel is 1.1×10^{-7} failures per hour. Estimate the probability of failure of the mechanical integrity of vessel protection layer.

9.7 A batch pharmaceutical production process has the potential for a runaway reaction. The runaway reaction will result in a process pressure far in excess of the design limitation of the reactor vessel. A runaway reaction will occur as the result of failure of the cooling water pump, which occurs with a frequency of 4.5×10^{-4} failures per hour. The vessel is protected against overpressure by a rupture disk and a relief valve. The rupture disk and relief valve are properly sized and piped to the vessel using separate connections. The failure probability of the rupture disk was estimated at 0.1%, and the failure rate of the relief valve, based on plant data, is 4.5×10^{-6} failures per hour. In addition, an operator who is continuously present has been trained to open a remotely operated manual dump valve when a high-pressure alarm is activated. The high-pressure alarm is activated 11 to 13 minutes prior to catastrophic rupture of the vessel as a result of the runaway reaction. Estimate the frequency at which a vessel rupture will occur for this scenario.

9.8 An operator has the ability to shut down a process heater if flame is lost because of contamination in the fuel supply. The operator is constantly monitoring the process and is relieved for breaks. In addition, the heater is equipped with flame detectors that sound an alarm in the event of flame loss. Using the simplified operator response analysis presented in table 9.1, estimate the operator's probability of failing to perform the shutdown function when required.

9.9 A catalyst production process generates a toxic gas during a reaction phase. The catalyst is generated using a batch process that has five steps: add reactants, mix, dry, react, and cool. The process operates continuously, with each batch requiring 10 hours to process. The drying phase requires two hours, and the reaction phase requires 90 minutes. If you were performing a LOPA for a SIF that would prevent the toxic gas from being released into the atmosphere, what use factor should you apply?

9.8 References

1. Dowell, A. M., III. "Layer of Protection Analysis for Determining Safety Integrity Level," *ISA Transactions* 37 (1998): 155-166.

2. Dowell A. M., III. "Layer of Protection Analysis and Inherently Safety Processes," *Process Safety Progress* 18, no. 4 (1999).

3. Gertman, David I., and Harold S. Blackman. *Human Reliability and Safety Analysis Data Handbook.* New York: Wiley-Interscience, 1994.

4. Lees, F. P. *Loss Prevention for the Process Industries.* London: Butterworth and Heinemann, 1992.

5. *Offshore Reliability Data Handbook,* 2d ed. Hovik: Det Norske Veritas, 1992.

CHAPTER 10

SIL Assignment

The objective of the safety integrity level (SIL) selection process—and of the entire practice of risk management—is to reduce the risk of the process under control to a tolerable level. The tolerable risk guidelines an organization selects must provide concrete guidance to the engineers making decisions at the safety systems level, as we discussed in chapter 3. The guidelines must allow the engineer to make a quantitative assignment of the amount of risk reduction that is required. After all, the SIL number the organization selects demands a real level of risk reduction as part of the safety requirements specification, since it defines the range of probability of failure on demand that the safety instrumented function (SIF) must provide. The final assignment of the SIL is where the theory of risk meets the reality of equipment design.

Although the assignment must be quantitative (i.e., a specific SIL category), the criteria can be either qualitative or quantitative. Organizations have developed a number of tools to help engineers convert unmitigated process risk to a required SIL before designing the safety system. These tools range from qualitative matrices and graphs all the way to quantitative guidelines on what is a tolerable risk of individual fatality. This chapter discusses four sets of guidelines for SIL assignment, which form the basis of essentially all SIL assignment protocols in use today.

The two qualitative guidelines covered in this chapter are *risk matrix* and *risk graph*. Risk matrices assign SILs based on the categories of the consequence and likelihood components of a risk. Risk graphs are also category based, but they directly include provisions for items such as occupancy and ability to escape, where other methods require these to be indirectly incorporated into the consequence analysis. In this chapter we will discuss how these matrix and graph processes are used in a purely qualitative sense, as well as how they can be calibrated to provide numerical ranges for the categories.

The other two guidelines are quantitative, since the clarity of numerical risk targets is preferred by some organizations and in some cases is required by law. Numerical risk targets make possible a straightforward quantitative determination of the frequency reduction required for a particular scenario. This number can then be converted into an SIF probability of failure on demand (PFD), and subsequently into the required SIL. The commonly used numerical risk targets described in this chapter include event-based frequency targets and event-based individual risk of fatality targets.

It is important for an organization to make consistent decisions about risk reduction. If decision-making appears to be ad hoc or process stakeholders consider the distribution of risks to be inequitable, the result may well be low morale, poor distribution of risk reduction resources, and other difficulties.

10.1 Correlating Required Risk Reduction and SIL

The SIL defines a quantitative category of performance for an SIF. Which metric is used to define performance depends on how often the SIF is expected to perform its action. The IEC 61508 and 61511 standards define two modes of operation: *demand mode* (also called "low demand mode") and *continuous mode* (also called "high demand mode"). The difference between the two modes is how often the safety system is called on to act.

Regardless of the mode of operation, each additional SIL number represents an additional order-of-magnitude reduction in risk. This relationship is easiest to remember for demand mode, since the SIL level corresponds exactly to the number of orders of magnitude of risk reduction. Thus, an SIL 2 SIF decreases the frequency of an accident by two orders of magnitude. Since the risk is the product of consequence and likelihood, this also represents a two order-of-magnitude reduction in overall risk.

The ANSI/ISA-84.01-1996 standard defines three SIL levels, with SIL 1 representing the least risk reduction and SIL 3 the most risk reduction (in agreement with table 10.1). The IEC 61508 and 61511 standards define four levels (shown in tables 10.1 and 10.2). The first three levels identically match the ANSI/ISA-84.01-1996 definitions, but the SIL 4 is added to designate a risk reduction of one order of magnitude more than SIL 3.

10.1.1 Demand Mode SIL Assignment

In the demand or low-demand mode of operation, demands to activate the SIF are infrequent compared to the test interval of the SIF. The IEC standards also define this mode for situations where the demands to activate the SIF are less than once per year. The demand mode of operation is the most common mode in the process industries and the only mode of operation considered in the ANSI/ISA-84.01-1996 standard. When defining SIL for the low-demand mode, you measure an SIF's performance in terms of average probability of failure on demand (PFD_{avg}). In this demand mode, the frequency of the initiating event, modified by the SIF's probability of failure, determines the frequency of unwanted accidents. We will describe the continuous or high-demand mode of operation in the next subsection (10.1.2). However, most of the remainder of this section (10.1), as well as the bulk of the earlier chapters of the text, focus on the demand mode of operation because of its prevalence in the process industries.

Table 10.1 shows the definitions of SIL levels for the demand mode of operation. The categories are described both in terms of PFD_{avg} and risk reduction factor. The risk reduction factor is the inverse of PFD_{avg}.

Table 10.1 Safety Integrity Levels: Demand Mode of Operation

DEMAND MODE OF OPERATION		
Safety Integrity Level (SIL)	**Average Probability of Failure on Demand**	**Risk Reduction Factor**
4	10^{-4} to 10^{-5}	10,000 to 100,000
3	10^{-3} to 10^{-4}	1,000 to 10,000
2	10^{-2} to 10^{-3}	100 to 1,000
1	10^{-1} to 10^{-2}	10 to 100

10.1.2 Continuous Mode SIL Assignment

IEC 61508 and IEC 61511 also define a continuous or high-demand mode of operation. In this mode of operation, demands are placed on the SIF much more frequently. The continuous mode is common in the machine industry and in avionics. In it, the frequency of an unwanted accident is determined by the frequency of an SIF failure. When the SIF fails, the demand for its action will occur in a much shorter time frame than the function test, so to speak of its failure probability is misleading. Essentially all of the dangerous faults of an SIF in continuous mode service will be revealed by a process demand instead of a function test. The IEC 61511 standard defines an SIF as operating in the continuous mode if the demand rate is higher than twice the proof test, with this definition being specific to the process sector. The IEC 61508 standard also includes systems in the continuous or high-demand mode where the demand rate is more than once per year.

Table 10.2 presents the definition of SIL levels for the continuous mode of operation. The categories are described in terms of frequency of dangerous failure per hour, with SIL 1 corresponding roughly to one dangerous failure every ten years or less.

Table 10.2 Safety Integrity Levels: Continuous Mode of Operation

CONTINUOUS MODE OF OPERATION	
Safety Integrity Level (SIL)	**Frequency of Dangerous Failure Per Hour**
4	10^{-8} to 10^{-9}
3	10^{-7} to 10^{-8}
2	10^{-6} to 10^{-7}
1	10^{-5} to 10^{-6}

10.1.3 SIL Assignment Equation: Demand Mode of Operation

As we noted earlier in section 10.1, SIL assignments are essentially order-of-magnitude categories of probability of failure on demand. In addition, the SIL category defined as SIL 1 has an upper limit of $PFD=10^{-1}$. On closer observation, one will soon note that the SIL category number is always the negative of the log of the upper limit of the SIL's PFD range. Although the standards organizations do not define SIL categories as such, a result of this definition is that we can calculate an SIL number given the PFD_{avg} by using equation 10.1.

$$SIL = -\log_{10}(PFD_{avg}) \tag{10.1}$$

Example 10.1

Problem: An SIF is used in a service that has a demand placed on it every two months. The proof-test interval of this SIF is once every year. The SIF was determined to have a dangerous failure rate of 4.5×10^{-8} per hour. What SIL category describes this SIF?

Solution: Since the demand rate is greater than twice the proof-test interval, the system is used in continuous mode. Perform continuous mode SIL assignment according to table 10.2. The dangerous failure rate of 4.5×10^{-8} per hour falls in between 1×10^{-7} and 1×10^{-8} per hour, so the SIL associated with the given failure rate is SIL 3.

Example 10.2

Problem: An SIF is used in the demand mode of operation. Quantitative calculations show that the SIF achieves a PFD_{avg} of 0.017. Assign an SIL to this SIF using both the SIL category tables and equation 10.1.

Solution: Since the demand mode of operation has been specified, use table 10.1 to make the SIL assignment. The PFD_{avg} of 0.017, or 1.7×10^{-2}, falls in between 1×10^{-1} and 1×10^{-2}, so the SIL associated with the given failure rate is SIL 1.

When equation 10.1 is used to calculate SIL the result is as follows:

$$SIL = -\log_{10}(0.017) = 1.77$$

The fractional component of the SIL that is shown when you use equation 10.1 provides information that describes how close the performance of a SIF is to another SIL category. As previously shown, a PFD_{avg} of 0.017 yields an SIL of 1.77, which falls into the SIL 1 category but is relatively close to SIL 2.

It is important to note that the standards only specify SIL categories as whole numbers. Moreover, they do not refer to SIL in such a way that it can be considered a real number (including a fractional component) as calculated using equation 10.1. Even so, many companies have found that using fractional SIL levels or specifying risk reduction factors in addition to the standard SIL categories is a useful way to increase SIF reliability.

10.1.4 Qualitative and Quantitative Assignment Methods

The assignment of an SIL, as well as the underlying risk tolerance criteria, can be considered either qualitatively or quantitatively. In the qualitative process, the parameters upon which a decision is based are subjective and are determined using experience and judgment. Qualitative SIL assignment methods require the analyst or team to select categories that descriptively identify the key risk parameters instead of selecting numeric values. An SIL selection then results from the qualitative categories that the team selects. It is also common to calibrate a qualitative SIL selection tool to fit quantitative risk parameters by setting numerical upper and lower bounds for the parameter categories. We discuss this in greater detail in section 10.7.

On the other hand, with quantitative methods we estimate the parameters that describe risk by calculation. These methods may also require establishing quantitative tolerable risk targets such as maximum individual risk of fatality. When using quantitative methods, a numerical value for risk is first calculated and compared to a quantitative target. The difference between the actual risk and the target risk is then calculated in terms of the PFD_{avg} of the SIF that would be required to move the actual risk to a tolerable level. This PFD_{avg} is then converted to the required SIL level.

In either case, the process results in the selection of SIL, which is always a quantitative description of the performance of an SIF. Because SIL is a quantitative target, it is possible to infer quantitative degrees of fault tolerance even when you have used qualitative methods for the selection process. In light of this, it is very important that the organization ensures that the SIL assignment methods and tools it uses will yield consistent results.

10.2 Hazard Matrix

The hazard matrix method of assigning SIL is one of the most popular assignment methods, especially in North America, because it is straightforward and can be applied easily. The hazard matrix method is qualitative and category based. The user must create a matrix that assigns broad categories to the consequence and likelihood components of the risk. For instance, the user could select consequences as "minor," "serious," or "extensive" and select likelihood as "low," "moderate," or "high." In some cases, the analyst uses quantitative tools, such as LOPA, to make it easier to determine which category to use. However, often the analyst does the assignment completely qualitatively, using his or her engineering judgment.

The consequence and likelihood each form an axis of the matrix, while each box of the matrix contains an SIL assignment. The analyst determines which box of the matrix corresponds to the selected categories of consequence and likelihood and selects the SIL in that box to address the risk under consideration. This SIL represents the amount of risk reduction required to move an event that has the selected consequence and likelihood to the tolerable risk region. The tolerable level of risk is implied by the structure of the matrix based on which SIL is in which box.

Table 10.3 shows three typical categories of consequence that analysts could use for one axis of the matrix, as described in the textbook *Guidelines for Safe Automation of Chemical Processes* from the AIChE's Center for Chemical Process Safety. Although this table shows three categories, matrices with four and five consequence categories are also common. The table contains a qualitative term for a consequence (or severity) rating, as well as a verbal description of the type of consequence typical of the category.

Table 10.3 Typical Consequence Categories: Hazard Matrix Example

Severity Category	Description
Minor	Impact initially limited to local area of the event with potential for broader consequence if corrective action is not taken.
Serious	One that could cause any serious injury or fatality on site or off site, or property damage of $1 million off site or $5 million on site.
Extensive	One that is more than 5 times worse than SERIOUS.

Note: Based on data in *Guidelines for Safe Automation of Chemical Processes.*

In this case, "minor" consequences are those that are initially limited to the area of the event. "Serious" consequences are those that could cause serious injury or fatality on or off site or could cause property damage between $1 million and $5 million. (Note that table 10.3 makes an effort to include consequences related to more than just personnel safety, such as property damage.) An "extensive" event is one that is five times worse than a "serious" accident.

You can select the proper category either qualitatively, using expert judgment, or with the help of quantitative calculation tools. The output from a quantitative consequence analysis might be, for instance, PLL=0.1, that is, a probable loss of life of 0.1 fatalities for the incident in question. Using this criterion, the analyst might select "minor" or "serious" as the consequence category. A PLL of 0.1 indicates that there is a 10 percent chance of one fatality, making a fatality unlikely. Even when you perform precise calculations of consequence, assigning a consequence category will require some subjective judgment since category descriptions may not be clear in all cases. When you calibrate categories by assigning numerical ranges, the process of incorporating quantitative calculations into the decision process becomes easier.

Table 10.4 shows three typical likelihood categories, as described in the *Guidelines for Safe Automation of Chemical Processes* textbook; matrices with four and five likelihood categories are also common. As in table 10.3, this table contains a qualitative term for a likelihood rating and a verbal description of the type of frequency considered typical of the category. These categories are also calibrated with quantitative ranges for the likelihood of an unmitigated event. The quantitative ranges for categories are often excluded from consequence and likelihood tables, which gives the analyst the responsibility of selecting the proper category based on his or her own best judgment.

Table 10.4 Typical Likelihood Categories: Hazard Matrix Example

Likelihood Category	Frequency (per year)	Description
Low	$< 10^{-4}$	A failure or series of failures with a very low probability that is not expected to occur within the lifetime of the plant.
Moderate	10^{-2} to 10^{-4}	A failure or series of failures with a low probability that is not expected to occur within the lifetime of the plant.
High	$> 10^{-2}$	A failure can reasonably be expected within the lifetime of the plant.

Note: Based on data in *Guidelines for Safe Automation of Chemical Processes.*

In this case, a "low" likelihood is one in which the event is not expected to occur within the lifetime of the plant and has a very low probability. This likelihood category is assigned a frequency of less than 1×10^{-4} events per year. The "moderate" likelihood event is also not expected to occur within the lifetime of the plant, but its probability is considered low rather than very low. This likelihood category is assigned a frequency range between 1×10^{-2} and 1×10^{-4} events per year. The "high" likelihood event is a failure than can be expected to occur within the lifetime of the plant, and is assigned a frequency range greater than 1×10^{-2} per year. It is important to note that the frequency used here is the frequency at which the event would occur without considering the preventive effects of the SIF under consideration. The frequency should, however, consider the protection layers present in the process, such as those described in section 9.4, that exist separately from the SIF.

You can select the category using expert judgment or with a likelihood analysis technique such as LOPA. For example, if LOPA showed that the event's frequency was 3.4×10^{-3} per year, then you would select the moderate category for likelihood. With the matrix method, the final assignment of SIL is made using a two-dimensional matrix similar to the one shown in figure 10.1. Variations of this technique exist that use a third dimension to represent the layers of protection available to prevent an incident. However, since LOPA more accurately assesses the effectiveness of the independent layers of protection, we do not describe this type of matrix here. The matrix shown in figure 10.1 is based on information provided in section E of IEC draft standard 61511-3, which describes matrix-based SIL selection methods in some detail. The matrix in figure 10.1 contains three consequence and three likelihood categories, which correspond to the tables that were used to select them. Each consequence-likelihood pair has an associated SIL level that represents the amount of risk reduction required to make the given situation tolerable.

Figure 10.1 Typical Hazard Matrix

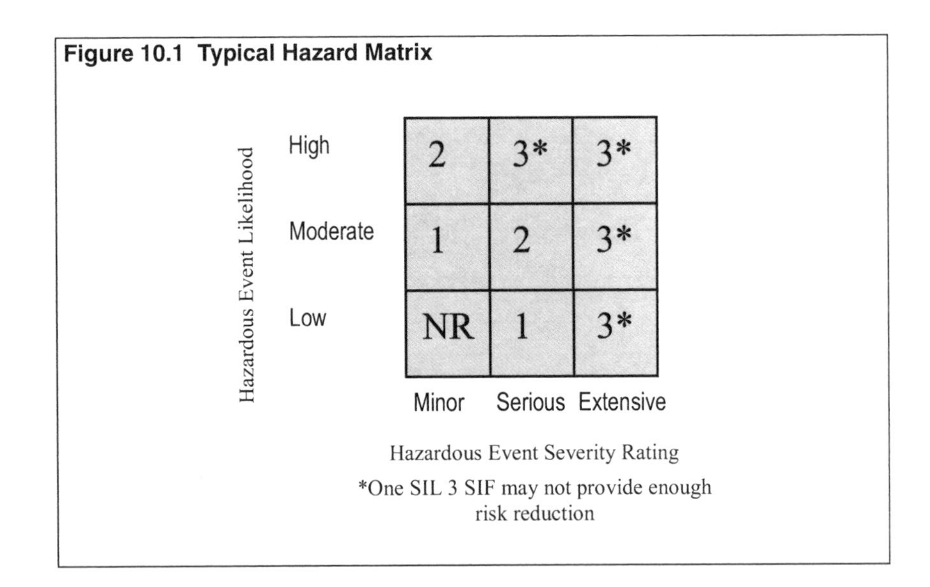

The matrix in figure 10.1 also includes some explanatory notes. First, an SIL assignment of NR indicates that for this situation, no risk reduction is required. The risk of the process is tolerable so no SIF need be installed. In this matrix, all of the SIL 3 boxes have an asterisk note that states, "One SIL 3 SIF may not provide enough risk reduction." If your analysis results in one of these boxes, you should consider more detailed analysis because an SIL 3 system may not be able to provide enough risk reduction for this situation. SIL 4 is not noted in this particular matrix because it is based on ANSI/ISA-84.01-1996, which does not recognize SIL 4. Although they do exist, such systems are often extremely difficult and expensive to design. Multiple independent systems with lower individual SILs or other more fundamental process changes are usually a more viable alternative.

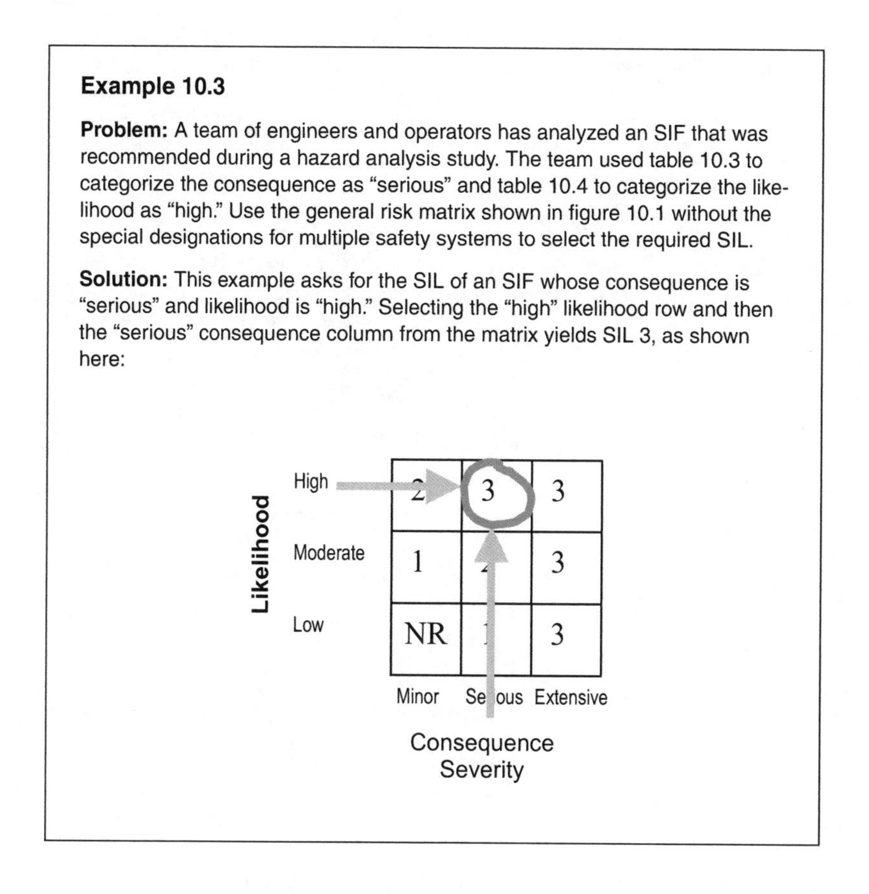

Example 10.3

Problem: A team of engineers and operators has analyzed an SIF that was recommended during a hazard analysis study. The team used table 10.3 to categorize the consequence as "serious" and table 10.4 to categorize the likelihood as "high." Use the general risk matrix shown in figure 10.1 without the special designations for multiple safety systems to select the required SIL.

Solution: This example asks for the SIL of an SIF whose consequence is "serious" and likelihood is "high." Selecting the "high" likelihood row and then the "serious" consequence column from the matrix yields SIL 3, as shown here:

10.3 Risk Graph

Risk graphs, like hazard matrices, are also qualitative and category based. Risk graphs were initially developed for German industry standards and are commonly used in the European Union. While a hazard matrix considers only likelihood and consequence, a risk graph considers likelihood, consequence, occupancy, and the probability of personnel avoiding the hazard. Each of these parameters is assigned a category. In some cases, quantitative tools, such as LOPA, are used to help the analyst determine which category to use, but typically the analyst makes the assignment qualitatively. Using the selected categories, the analyst follows the resulting path that leads to the box containing the SIL assignment. As with hazard matrices, the tolerable level of risk is implied both in the structure of the risk graph and by which SIL is at the end of each particular path. The risk graph methods described here are consistent with those in IEC

Example 10.4

Problem: The team from example 10.3 believes that an SIL 3 SIF is excessively conservative and a waste of valuable resources for this application. They elect instead to perform quantitative analyses to estimate more accurately the consequence and likelihood of the accident being prevented than is possible with the simple qualitative evaluation. This more detailed analysis estimates the consequence to be a PLL of 0.21 and the frequency to be once in 576 years. Using figure 10.1 as in example 10.*3*, what SIL should be selected?

Solution: This example asks for the SIL of an SIF whose consequence is a probable loss of life of 0.21 and a likelihood of 1/576 events per year. The PLL=0.21 fits into either the "minor" or "serious" category. For this example, "minor" is selected to represent a PLL of 0.21. A likelihood of 1/576 incidents per year falls into the 10^{-2} to 10^{-4} range, described as of "moderate" likelihood. Selecting the "minor" consequence column and then the "moderate" likelihood row from the matrix yields SIL 1, as shown here:

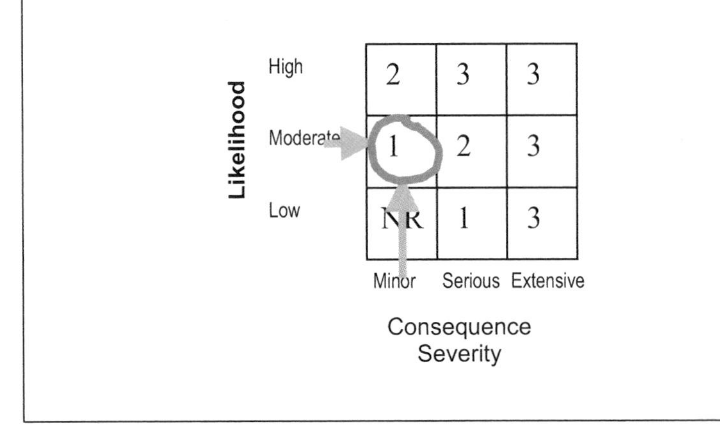

draft standard 61511, which are primarily based on dealing with the injury or death of people rather than other forms of harm.

Risk graph analysis uses four parameters to select the SIL: consequence, occupancy, probability of avoiding the hazard, and demand rate. The consequence parameter represents the average number of fatalities likely to result from a hazard when the area is occupied. The parameter should include the expected size of the hazard and the receptor's vulnerability to the hazard. Occupancy is a measure of the length of time that the impact area or effect zone of the incident outcome is occupied. The probability that the hazard can be avoided will depend on what methods the personnel have for knowing that a hazard exists and what means they have to escape the hazard. The demand rate is the likelihood that the accident will occur. This rate is arrived at without considering the effect of the SIF being studied, but including all other non-SIS protection layers.

Table 10.5 shows typical consequence categories, as described in section D of IEC draft standard 61511-3. The table shows four consequence categories, which is the number most frequently used in risk graphs, but is not a strict limitation. The table contains a two-letter code to describe each category as well as a verbal description or probable loss of life (PLL) range considered typical for that category.

Table 10.5 Typical Consequence Categories: Risk Graph Example

Category	Description
C_A	Minor Injury
C_B	PLL = 0.01 to 0.1
C_C	PLL = 0.1 to 1.0
C_D	PLL > 1.0

In table 10.5, consequences are based on the probable loss of life related to the accident scenario. If the PLL is less than 0.01, then the category is C_A. Likewise, a PLL from 0.01 to 0.1 relates to category C_B, a PLL from 0.1 to 1 relates to category C_C, and a PLL that is greater than 1 relates to category C_D. Other references, such as IEC 61508, part 5, contain consequence descriptions that are less probabilistic than the PLL-based assignments shown here. For instance, C_A could relate to minor injury, C_B to serious injury or fatality, C_C to multiple fatalities, and C_D to a large number of fatalities. Using table 10.5, an event that has a PLL of 0.2 will fall into the C_C category.

It is worth noting that the PLL designation can be seen to imply a probability in addition to the consequence magnitude. When using the PLL as a measure of consequence with a risk graph, take care to properly assign the other parameters of the incident so the consequence parameter is indeed considered independently. Some organizations find this procedure difficult to implement consistently, and thus prefer the latter set of consequence category designations.

Table 10.6 shows typical occupancy categories as described in section D of IEC draft standard 61511-3. The table only provides two categories of occupancy, which is typical of risk graph analysis. In the table, the two-letter identifiers for each occupancy category designate a fraction of time that the area that would be impacted by the hazard is occupied by personnel.

Table 10.6 Typical Occupancy Categories: Risk Graph Example

Category	Description
F_A	Rare to more frequent exposure in the effect zone of the incident. The effect zone is occupied less than 10% of the time.
F_B	Frequent to permanent exposure in the effect zone of the incident.

F_A is assigned to the occupancy parameter if the amount of time that the effect zone of the incident is occupied is less than 10 percent of the operating time. If the hazard zone is occupied more frequently, F_B is selected. A process unit in which there is a fifty-fifty chance that an operator will be in the hazard zone at any one time is assigned an occupancy class of F_B.

The analyst must be careful when selecting an occupancy category for a risk graph to ensure that credit for reduced occupancy is not taken in multiple parts of the analysis. For example, he or she may have already considered credit for occupancy in any one of several other components of the analysis. When an analyst is calculating the consequence of an incident quantitatively, he or she almost always considers the occupancy of an effect zone when determining the probable loss of life. In addition, the analyst may have considered the probability of occupancy of the effect zone as a mitigating event when performing a layer of protection analysis. If the occupancy was considered in either of these two steps, then no additional credit can be taken during this graph portion of the analysis. In the case where the analyst considered occupancy when determining either the likelihood or consequence, he or she must choose the F_B category, which reflects no credit for decreased occupancy, for the risk graph occupancy parameter.

Table 10.7 shows typical probability of avoidance categories, as described in section D of IEC draft standard 61511-3. This table shows two categories, which is typical of risk graph analysis. The two-letter identifiers for each probability of avoidance category indicate the probability that an exposed operator would be able to, first, detect a hazardous condition and, next, have a means of escaping the effects of that condition.

Table 10.7 Typical Probability of Avoidance Categories: Risk Graph Example

<table>
<tr><th>Category</th><th>Description</th><th>Conditions Allowing P_A</th></tr>
<tr><td>P_A</td><td>Selected if all conditions to the right are met (i.e., credit for probability of avoidance is taken).</td><td rowspan="2">P_A should only be selected if the following conditions are true:
• The operator will be alerted if the SIS has failed.
• Facilities are provided for avoiding the hazard that are separate from the SIS and that enable escape from the hazardous area.
• The time between operator's alert to a hazardous condition and the occurrence of the event is greater than one hour or is definitely sufficient for the necessary actions.</td></tr>
<tr><td>P_B</td><td>Conditions to the right are not satisfied. No credit for probability of avoidance is taken.</td></tr>
</table>

In essence, the analyst selects the probability of avoidance category by evaluating a checklist that will determine whether avoidance credit can be taken or not. If credit for probability of avoidance can be taken, the analyst selects category P_A. The checklist of items that must be true before credit can be taken for avoiding the hazard varies from organization to organization. As an example, P_A is assigned to the probability of avoidance parameter if all of the following conditions are true, based on the example in IEC 61511-3:

1. Facilities are provided to alert the operator that the SIS has failed. If a field operator who is exposed to a hazard does not know that a hazardous condition exists, he or she cannot reasonably be expected to avoid the hazard. In cases where an SIS may not be part of the process, this should be understood to mean that the operator is specifically alerted that the harmful outcome is imminent.

2. Independent facilities are provided for shutdown such that the hazard can be avoided or that enable persons to escape the area safely. In order to take credit for the operator's escape, the operator must, of course, have a way to make that escape possible. In deciding whether or not an operator may escape from an accident, the organization should consider requiring the operator to take action to shut down the process to avoid the hazard. The organization should also consider what means of escape from a hazardous area the operator has available.

3. The time between the moment the operator is alerted and the hazardous event occurs exceeds one hour or is definitely sufficient for the necessary actions. As we described in chapter 6, explosions travel at the speed of sound, and thermal radiation from fires travels at the speed of light. An operator in the effect zone will not be able to escape from such a hazard with any more probability than the vulnerability that is already included in the consequence calculation. When determining probability of avoidance, the analyst should consider the amount of time that elapses between the clear alert of the release incident and the moment the incident turns into a fire or explosion incident outcome case.

If all three of these conditions are not met, the analyst must select P_B. With category P_B, no credit is given for the possibility that persons in the effect zone of an incident outcome will be able to escape the hazard.

As when assigning credit for occupancy, the analyst must take care when assigning credit for probability of avoidance to ensure that credit is not taken in multiple parts of the analysis. When quantitatively calculating consequences of an incident outcome, the analyst should almost always include the probability of avoidance in the analysis. When calculating an effect zone, the analyst must select an endpoint for the analysis. For instance, pool fire effects are often calculated to an endpoint of 12.5 kW/m^2. It is assumed that all persons who are inside the zone that is calculated using this endpoint have a certain vulnerability to fatality. This vulnerability includes their attempts to escape from the hazard. Therefore, when calculating quantitative consequences, for instance in terms of probable loss of life (PLL), the analyst should include the probability of avoidance, and no additional credit should be taken in the risk graph. When the analyst includes credit for probability of avoidance in the consequence analysis, he or she should always include a probability of avoidance factor P_B.

Table 10.8 shows three typical demand rate categories, as described in section D of IEC draft standard 61511-3. Three categories is typical of risk graph analysis. The table shows two-character identifiers for each demand rate category that relate to the unmitigated frequency at which the hazardous event will occur.

Table 10.8 Typical Demand Rate Categories: Risk Graph Example

Category	Description
W_1	Less than 0.03 times per year.
W_2	Demand rate between 0.3 and 0.03 times per year.
W_3	Demand rate between 3 and 0.3 times per year. (NOTE: This risk graph is not valid for demand rates higher than 3 times per year.)

Table 10.8 assigns demand rate categories based on quantitative rate information. The demand rate described in this table is the frequency of the accident that would result if the process under control were not equipped with the SIF you are considering. However, it does include the effects of non-SIS layers of protection. This definition of demand rate is identical to the likelihood term we described for hazard matrices. If the demand rate is less than 0.03 per year, the category is W_1. Likewise, a demand rate from 0.03 to 0.3 per year relates to category W_2, and a demand rate from 0.3 to 3 per year corresponds to category W_3. A demand rate that is higher than three occurrences per year will require recalibration of the risk graph to account for high-frequency demands. Other reference sources contain demand rate descriptions that are more qualitative than the frequency-based assignments shown in table 10.8. In IEC 61508-5, for instance, W_1 relates to a very slight probability of the unwanted event occurring, W_2 relates to a slight probability of the unwanted event, and W_3 relates to a relatively high probability. Using table 10.8, we see that an event that has an unmitigated frequency of 0.9 times per year will fall into the W_3 category.

A combination of consequence, likelihood, occupancy, and probability of avoidance represents the level of unmitigated risk. Once you have determined those categories, use the risk graph to determine the SIL that will reduce the risk by the amount necessary to achieve a tolerable level of risk. Figure 10.2 contains a typical risk graph, as presented in IEC draft standard 61511-3. The SIL is selected by drawing a path from the starting point on the left to the boxes at the right by following the categories selected for consequence, occupancy, and probability of avoidance. The combination of those three determines which row is selected. Which specific box is selected within the selected row is determined by which demand rate category has been selected. This specific box will contain one of the following designations: 1, 2, 3, 4, a, b, or "---". If the box contains a numeral, that numeral is the SIL required for the SIF. If the box contains an "a," then no special requirements (e.g., an SIS) are required to achieve a tolerable level of risk. If the box contains a "b," then a single SIS alone is insufficient for reducing the risk to a tolerable level. If the box contains a "---," then no safety requirements exist at all for the function.

Figure 10.2 Typical Risk Graph

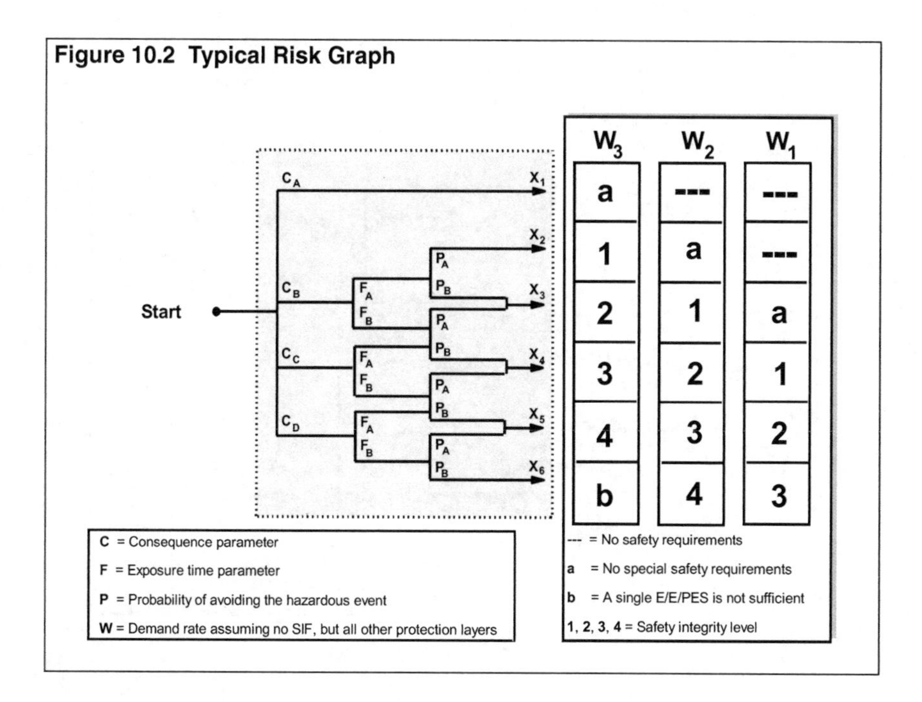

Consider an analysis of SIF risk that yielded C_B, F_B, P_B, and W_2. By following the first three terms through the risk graph, we arrive at row X_4. The demand rate W_2 leads to the selection of the second column in the X_4 row, which yields SIL 2.

Again, it is important to note that, as with hazard matrices, the SIL that is associated with each box on the risk graph implies a specific tolerable risk level. Before applying such methods, it is vital that the analyst confirm that the risk-level calibration of the tool is consistent with the tolerance of risk in the given situation.

Example 10.5

Problem: A hazard assessment considers a recommended safety instrumented function (SIF). The team determines that a single fatality is the most likely consequence of the incident that the SIF is preventing. Due to the configuration of the plant, the area is normally occupied, and there is no possibility of avoiding the hazard. The team also determines that the demand rate of the hazard is low, but not inconceivable. Using figure 10.2 and tables 10.5 through 10.8, what SIL should be selected?

Solution: Based on the description of consequence, the PLL is between 0.5 and 1.0, which falls into the C_C category. Since the area is normally occupied, choose F_B for occupancy. No possibility of avoidance requires that P_B be selected for probability of avoidance. A low demand rate falls into the W_2 category, which is between the very slight probability of W_1 and the relatively high probability of W_3. Following from the starting point using the selected categories yields row X_5. Since W_2 is selected, choose the middle column. The selected box contains SIL 3, as shown here:

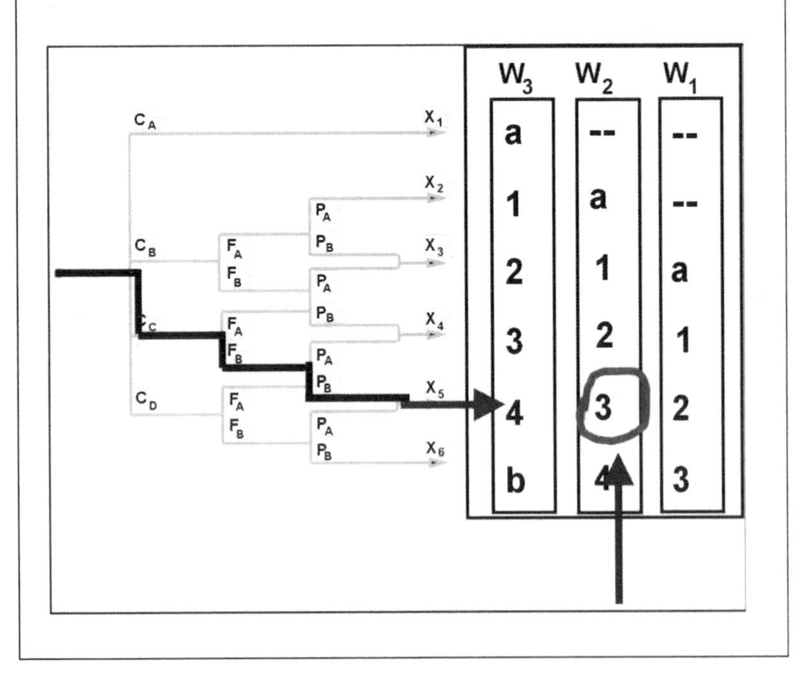

Example 10.6

Problem: The team from example 10.5 believes that an SIL 3 SIF is excessively conservative for this application. As a result, they elect to perform quantitative analyses to estimate the consequence and likelihood of the accident being prevented more accurately than the simple qualitative evaluation makes possible. The resulting quantitative estimate for the consequence is a PLL of 0.21 with a frequency of once in 576 years. Using figure 10.2 and tables 10.5 through 10.8, what SIL should be selected?

Solution: A PLL of 0.21 falls into the C_C category. The consequence analysis method that yields a probable loss of life (PLL) accounts inherently for the occupancy and probability of avoidance. As such, no additional credit should be taken in the risk graph analysis, so F_B is chosen for occupancy and P_B is selected for probability of avoidance. A demand rate of 1/576, or 0.0017, falls into the W_1 category. Following from the starting point using the selected categories yields row X_5. Since W_1 is selected, the far right column is chosen. The selected box contains SIL 2, as shown here:

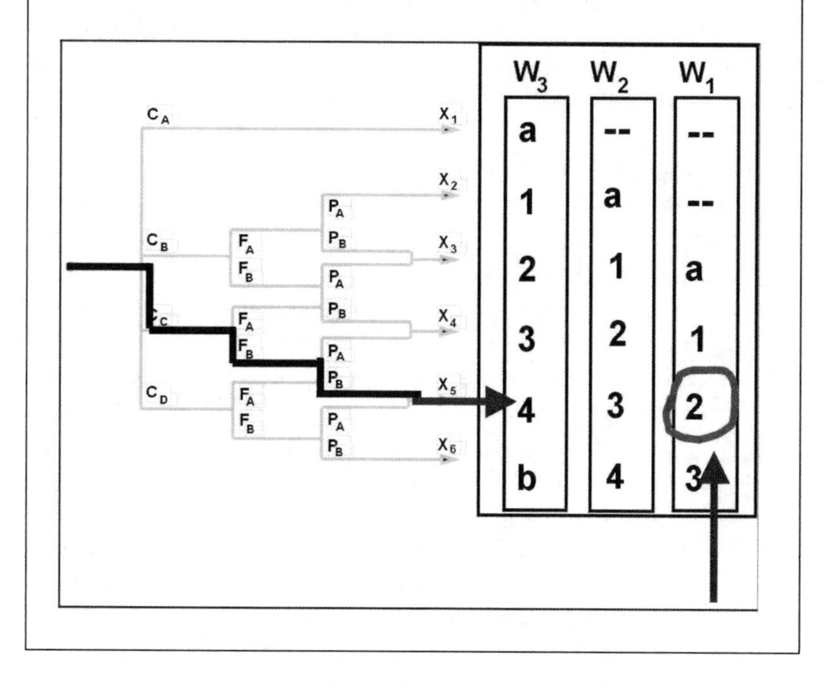

10.4 Incorporating LOPA into Qualitative Methods

Accurately accounting for the protection layers incorporated into the process is an important component of selecting an SIL. Layer of protection analysis (LOPA), as described in chapter 9, is a quantitative tool that allows the user to quickly and accurately calculate event frequencies based on initiating event frequencies and the probability of protection layers failing. We discussed incorporating quantitative LOPA in the earlier sections of this chapter on hazard matrices and risk graphs. These sections explained that you can use quantitative LOPA to calculate an event frequency, which you can then use to select the likelihood (or demand rate) category more accurately than is possible using expert judgment alone.

Some organizations have found that incorporating qualitative LOPA into their SIL selection process has enabled them to improve their results with minimal effort. Qualitative LOPA is more concerned with determining the number of available protection layers than it is with the actual probability of failure of each of those layers. We then use the total number of protection layers to modify either the likelihood category or the final SIL selected from the tool.

When incorporating qualitative LOPA into a hazard matrix or risk graph, the selection process is performed in the same way for all of the variables, with the exception of likelihood (or demand rate). Typically, the likelihood category is selected based on the overall likelihood of the incident you are studying, including the initiating events, available protection layers, and mitigating events. When incorporating qualitative LOPA into this process, select the likelihood category based only on the initiating event(s). At this stage, we do not consider the effect of protection layers.

After you have selected the risk categories, considering only the initiating event(s), you determine the available layers of protection. When performing a qualitative LOPA, it is important to adhere to a strict definition of what constitutes a protection layer. The qualifications used to define a formal protection layer are specificity, independence, dependability, and auditability, as described in section 9.2. The dependability component of a protection layer assumes a greater importance in qualitative LOPA, where each protection layer is effectively assigned the capability of reducing risk by one order of magnitude. As such, it is critical in this form of analysis to ensure that the probability of failure of a protection layer is less than 10 percent.

After determining the number of available protection layers, use this figure in the analysis to either modify the event likelihood category or the selected SIL. The action you choose will depend on the preferences of your organization, but the results will be roughly equivalent.

Consider the demand rate categories in table 10.8. If the SIL selection team determined that the initiating event frequency yielded a category of W_3, and one protection layer was available to prevent the incident, the overall likelihood category could then be changed to W_2 by incorporating the effect of the one protection layer. If there were two protection layers available in this scenario, the selected likelihood category could be decreased to W_1. Similarly, if the W_3 category selected for the initiating event yielded an SIL 4 when the other risk parameters were considered, the overall required SIL would be decreased to SIL 3 by including the one protection layer in the analysis.

Example 10.7

Problem: A team is incorporating qualitative LOPA into their analysis. The team is trying to determine if their ignition control program, which limits the sources of ignition available for igniting a release, should be considered a protection layer. Statistics for the process show that when the event they are trying to prevent occurs—the release of hydrocarbon gas—there is a 30 percent chance that ignition will occur. Should the team consider ignition control as a layer of protection?

Solution: No.

A layer of protection must meet the four criteria of specificity, independence, dependability, and auditability. In addition, to use qualitative LOPA each protection layer must provide at least one order of magnitude of risk reduction. While one could argue that the specificity, independence, and auditability attributes of an ignition control program are present for this protection layer, the required dependability is not. Since statistics show that 30 percent of releases are ignited, the ignition control "protection layer" has a failure probability of 30 percent, which does not provide the required one order-of-magnitude risk reduction for the qualitative method.

10.5 Assignment Based on Frequency

The method of using frequency-based risk targets is inherently quantitative. The analyst selects a maximum allowable frequency target based on the consequence of the hazard that the SIF in question acts to prevent. The required risk reduction is the difference between the unmitigated event frequency and the maximum event frequency target for the consequence being considered. The selected SIL represents the probability category for SIS failure on demand that will ensure that the mitigated event frequency does not violate the maximum frequency target. As we noted previously, this SIL corresponds to the required risk reduction.

When using frequency-based targets, tolerable frequencies for unwanted accidents must be established. The frequency that is tolerable in each case will depend on the consequence of the event. Table 10.9 provides some typical tolerable frequency limits for a range of consequences, based on information in draft IEC 61511. The table includes a description of a consequence category and the associated maximum allowable frequency target for each consequence.

According to table 10.9, a target frequency of no more than 1×10^{-4} per year is specified for "minor" consequences. A target frequency of no more than 1×10^{-5} per year is assigned to "serious" consequences, and 1×10^{-7} per year is the maximum assigned for "extensive" consequences. The precise definitions of "minor," "serious," and "extensive" and the three frequencies they are associated with thus correspond to the tolerable level of risk in the situation under consideration.

Example 10.8

Problem: An SIL selection team is using a risk graph method supplemented by qualitative LOPA to modify the required SIL. They have determined that, for the event they are studying, the consequence would be multiple fatalities, with no possibility of avoidance and no credit taken for lack of occupancy. The initiating event in this scenario occurs with a frequency of once in ten years, and there are two protection layers that prevent the initiating event from propagating into an incident. Using figure 10.2 and tables 10.5 through 10.8, what SIL should be selected?

Solution: An incident with a multiple fatality consequence falls into the C_D category. Based on the description of the scenario, the team should choose F_B for occupancy and P_B for probability of avoidance. A demand rate for the initiating event of 1/10, or 0.1, falls into the W_2 category. Following from the starting point using the selected categories yields row X_6. Since W_2 is selected, the team should choose the far middle column. The selected box contains SIL 4, as shown here:

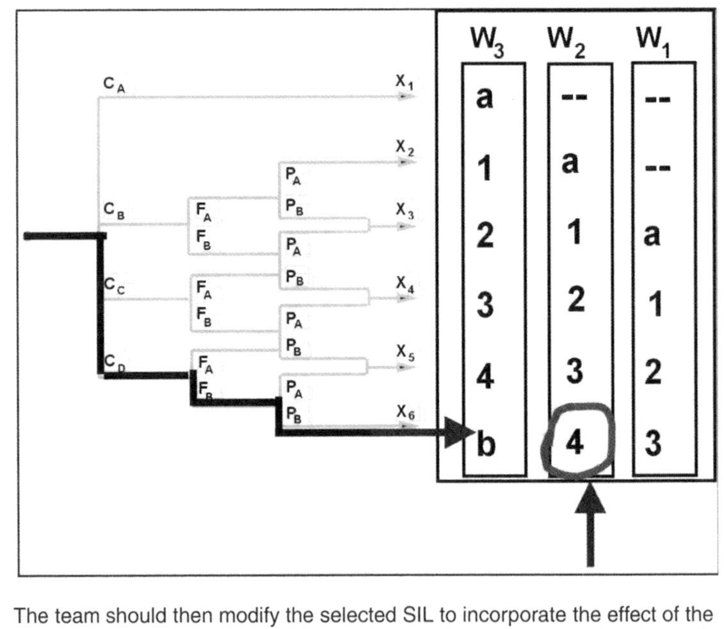

The team should then modify the selected SIL to incorporate the effect of the independent protection layers. For each protection layer, one order of magnitude of risk reduction is assumed. This means that the selected SIL (i.e., 4) can be decreased by 2, yielding a required SIL of 2.

Table 10.9 Typical Tolerable Frequency Categories

Category	Tolerable Frequency
Minor	1×10^{-4} per year
Serious	1×10^{-5} per year
Extensive	1×10^{-7} per year

The amount of risk reduction that an SIF can provide is a function of how much the SIF can decrease the frequency of an unwanted event. The decrease in frequency of the unwanted event is a function of the PFD of the SIF. As described in table 10.2, SILs are order-of-magnitude categories of PFD for demand-mode operation.

For an event mitigated by an SIF to occur anyway, both the unmitigated event must occur and the SIS must fail. Since the events are logically related by an AND, the resulting mitigated event frequency is calculated using probability multiplication. In this instance, the PFD of the SIF is the result to be determined, so the probability multiplication equation is solved for PFD, as shown in equation 10.2. A complete derivation of equation 10.2 is included in appendix A.

$$PFD_{SIF} = \frac{f_{Tolerable}}{f_{Event,\ No\ SIS}} \qquad (10.2)$$

After calculating the probability of failure on demand (PFD) that is required to achieve the tolerable frequency, convert that PFD (or risk reduction factor—RRF) into an SIL category for that SIF. For instance, if SIL 2 is assigned to an SIF, then the RRF of the system must be between 100 and 1,000. When you assign this SIL 2, a risk reduction factor of 100 is the best level of performance that can safely be assumed for the system. Because you can only assume the least effective performance in the RRF range, the SIL you select must be one number higher than the RRF range into which the calculated PFD falls.

Consider the case where a PFD of 5.0×10^{-3} (i.e., RRF = 200) is required to meet tolerable risk goals. The PFD falls into the range of 10^{-2} to 10^{-3} represented by SIL 2 (i.e., RRF of 100 to 1,000). In this case, SIL 3 will be required. SIL 2 is insufficient because every number and thus every subsequently designed system in the SIL 2 range will *not* meet the tolerable risk goals. If SIL 2 is specified, then an engineer can supply a system with a PFD of 9.0×10^{-3} (RRF = 111). Even though this number falls into the SIL 2 range, the system will not meet the tolerable risk targets. To ensure that risk targets are met, you will need an SIL that is one number greater than the range of the target PFD if the SIL number alone is the only integrity target you are using to specify the performance of the system.

Many end users have begun to provide more information in their safety requirements specifications than just the SIL alone. This is because of the implications of overspecifying SIL from the broad, order-of-magnitude categories. According to the various standards of practice, the SIL category that is selected for an SIF has greater implications than the order-of-magnitude category of PFD alone. These additional criteria that are assigned based on the SIL include the level of dangerous fault tolerance required by all of the subsystems of an SIF, as well as the methods and degree of validation and verification that you must use when programming an SIF. Since specifying a higher-than-necessary SIL has more significant ramifications than simply increasing the PFD of the SIF, many users have elected to specify both an SIL category and a target risk reduction factor, or a fractional SIL. For instance, if a function required an RRF of 35, the traditional approach would have been to select an SIL 2. Some users, however, elect to specify SIL 1 with a risk reduction factor of 35, or an SIL of 1.54 instead. This type of specification will allow qualitative aspects of SIL 1 to be used, while requiring a higher level of performance than a risk reduction factor of 10. That risk reduction factor still could be provided only if SIL 1 were specified. This procedure thus takes maximum advantage of the range of SIL performance requirements present in the standards.

Example 10.9

Problem: A hazard assessment team considers a recommended safety instrumented function (SIF). It has performed quantitative analysis of the risk, yielding a consequence of PLL=0.21 with a likelihood of 1/576 events per year. The team decides that a PLL of 0.21 equates to a "minor" consequence in their tolerable event frequency table, as shown in table 10.9. Using both an SIL-only assignment and an SIL-plus-RRF assignment, select the most appropriate safety integrity level (SIL) specification for this situation.

Solution: Based on the description of the consequence, the category selected is "minor," which has an associated maximum target frequency of 1.0×10^{-4} per year. The target frequency of 1.0×10^{-4} and the unmitigated event frequency of 1/576 are combined using equation 10.2 to calculate the maximum PFD. The result is a required PFD of 0.058, which has an associated risk reduction factor of 17, as shown here:

$$PFD_{SIF} = \frac{f_{Tolerable}}{f_{Event,\ No\ SIS}} = \frac{1 \times 10^{-4}}{1/576} = 0.058$$

$$RRF = \frac{1}{0.058} = 17$$

A probability of failure on demand (PFD) of 0.058 (RRF=17) falls into the range of SIL 1. However, if only SIL 1 were specified, a system with an RRF of 10 could be supplied, leaving the risk reduction goals unattained. As a result, the team must specify SIL 2 if SIL is the only performance target specified. As an option, the team may specify SIL 1 with a risk reduction factor of 17 as an alternate set of requirements.

10.6 Assignment Based on Individual and Societal Risk

Of the SIL assignment methods we discuss in this book, individual risk-based targets are the most quantitative technique. We calculate the maximum allowable frequency target based on the maximum allowable individual risk for an event and on the probable loss of life resulting from that event. In addition, we can also adjust the maximum tolerable risk of fatality by a risk aversion factor to weight multiple fatality events more heavily. As with the frequency-based target method, the required risk reduction is the difference between the unmitigated event frequency and the event frequency target that was calculated for the scenario.

If you use individual risk criteria, then you can calculate a tolerable frequency for the event based on the probable loss of life caused by the event, the maximum individual risk target for an event, and a risk aversion factor. This risk aversion factor is often designated as "α" and provides extra weighting to events that have a higher loss of life. This corresponds to the perception that those events require significant additional risk reduction. We then calculate the target frequency as the individual risk of fatality frequency divided by the probable loss of life taken to the power of the risk aversion. This process is related to that described in chapter 3, "Tolerable Risk." Once you have determined the frequency target, you can calculate the risk reduction or required PFD using equation 10.3:

$$f_{Tolerable} = \frac{f_{Tolerable,\ Individual}}{PLL^{\alpha}} \qquad (10.3)$$

10.7 Calibrating Hazard Matrices and Risk Graphs

As we showed in sections 10.2 and 10.3, the categories that comprise risk graphs and hazard matrices either can be entirely qualitative or can contain ranges of quantitative values. In either case, the assigned SIL is a quantitative description of the amount of risk reduction required to make the risk tolerable. Since the level of required risk reduction is always given as a quantitative value, we can back-calculate the tolerable levels of risk. If we know both the accident frequency without the SIF and the consequence, then we can determine the tolerable frequency of that accident and even the tolerable individual risk of fatality using equations 10.2 and 10.3. This back-calculation thus determines the calibration of the matrix or graph being used.

Consider the risk graph shown in figure 10.1. In the case where the consequence is considered "serious" and the likelihood is considered "moderate," the risk graph dictates that SIL 2 risk reduction is required. Based on the category descriptions given in tables 10.3 and 10.4, a pessimistic view of the "serious" consequence category would be a PLL of less than 1.0. Similarly, a pessimistic view of the "moderate" likelihood category would be a frequency of less than 1.0×10^{-2}. Based on the SIL definitions for demand mode in table 10.1, a pessimistic view of the PFD for

Example 10.10

Problem: A hazard assessment team considers two recommended safety instrumented functions (SIFs). They have performed quantitative analysis of the risk, yielding a consequence of PLL=0.21 for the first event and a consequence of PLL=2.5 for the second event. A LOPA yielded a likelihood of 1/576 events per year for both events. The facility for which this SIF is being considered has a maximum individual risk of fatality criterion of 2.0×10^{-5} and uses "risk-averse" societal risk criteria where the risk aversion factor is 2. Using an SIL-only assignment, an SIL-plus-RRF assignment, and a "fractional" SIL assignment according to equation 10.1, select the most appropriate safety integrity level specification for this situation.

Solution: The first step in assigning an SIL when the tolerable risk criteria are expressed in terms of maximum individual and societal risk is to determine the tolerable frequency of the specific event such that overall risk targets are not exceeded. The team should determine the tolerable event frequency using equation 10.3. The maximum individual risk of fatality at this facility is 2.0×10^{-5}, and the risk aversion factor (α) is 2. Using equation 10.3 for the case where the event's PLL is 2.5 yields a tolerable frequency of 3.2×10^{-6}:

$$f_{Tolerable} = \frac{f_{Tolerable,\ Individual}}{PLL^{\alpha}} = \frac{2.0 \times 10^{-5}}{2.5^{2}} = 3.2 \times 10^{-6}$$

When the PLL of an event is less than 1, however, the risk aversion factor should not be used and the team should set α to be 1. In this case, the PLL reflects a 0.21 probability of a single fatality occurring. Therefore, the tolerable frequency of the PLL=0.21 event is 9.5×10^{-5}:

$$f_{Tolerable} = \frac{f_{Tolerable,\ Individual}}{PLL^{\alpha}} = \frac{2.0 \times 10^{-5}}{0.21^{1}} = 9.5 \times 10^{-5}$$

Now that the tolerable event frequencies are known, the team combines them with the event frequencies before applying the SIF to calculate the required PFD using equation 10.2. The result is a required PFD of 5.5×10^{-2} for the PLL=0.21 case and a PFD of 1.8×10^{-3} for the PLL=2.5 case. These cases have associated risk reduction factors of 18 and 560 respectively. Note that both RRFs are rounded to two significant figures, based on the inherent accuracy of these types of calculations.

$$PFD_{SIF} = \frac{f_{Tolerable}}{f_{Event,\ No\ SIS}} = \frac{9.5 \times 10^{-5}}{1/576} = 5.5 \times 10^{-2}$$

$$RRF = \frac{1}{0.055} = 18$$

$$PFD_{SIF} = \frac{f_{Tolerable}}{f_{Event,\ No\ SIS}} = \frac{3.2 \times 10^{-6}}{1/576} = 1.8 \times 10^{-3}$$

$$RRF = \frac{1}{0.0018} = 560$$

The PLL=0.21 case requires either SIL 2, SIL 1 with RRF of 18, or SIL 1.3, depending on the facility's method for assigning SILs. Similarly, the PLL=2.5 case will require either SIL 3, SIL 2 with RRF of 560, or SIL 2.7.

this system would be less than 1.0×10^{-2}. Using these figures for safety system PFD and unmitigated event frequency as boundaries, it is then possible to calculate the maximum tolerable event frequency. To do so, we would use a rearranged version of equation 10.2, as shown in equation 10.4:

$$f_{Tolerable} = PFD_{SIF} \times f_{Event,\ No\ SIS} \quad (10.4)$$

In this case, the tolerable frequency is 1.0×10^{-4}. We can use a similar approach with the values for the tolerable event frequency and the PLL of that event. That is, we can calculate the maximum individual risk by rearranging equation 10.3, as shown in equation 10.5:

$$f_{Tolerable,\ Individual} = PLL^{\alpha} \times f_{Tolerable,\ Event} \quad (10.5)$$

When using equation 10.5, we must make some assumptions about risk aversion if the PLL of the incident is greater than one. If we are analyzing an entire matrix, as opposed to a single assignment, we can determine the risk aversion factor by varying its value by trial and error until all of the boxes in the risk graph are consistent. Otherwise, there will not be enough information to calculate all of the parameters for the case where the PLL is greater than one. For the PLL=1 scenario we are analyzing here, the maximum tolerable individual risk of fatality is 1.0×10^{-4} per year.

As this example demonstrates, it is a relatively straightforward exercise to reverse engineer the maximum tolerable individual risk of fatality that underlies the qualitative tools of the risk graph and hazard matrix. It is also interesting to note that in the example, the consequence was not even described in terms of a quantitative range. It was simply a qualitative description from which a quantitative value was inferred. Although we performed this example for a hazard matrix, the same type of analysis could easily be performed for a risk graph.

To make qualitative tools for SIL assignment internally consistent from box to box and also consistent with an organization's overall safety goals, these tools must be properly calibrated. The process of assigning an SIL to the boxes in a risk graph or hazard matrix—each of which represents some level of process risk—is called "calibrating" the tool. We perform the calibration by calculating the risk posed by the potential events represented by each box in the tool. Next, we analyze an organization's risk tolerance on a quantitative basis, and then use equations 10.3 and 10.2 to assign the SIL that will reduce the risk to a tolerable level. SIL is typically assigned by performing a minimum value calibration. This determines the SIL required in the most pessimistic case for a range of data. Some organizations also use a prototypical calibration process in which each category of consequence and likelihood is represented by a sample value, which is selected to represent all values in the range. The calibration is then performed using this single prototypical value.

10.7.1 Minimum Value Calibration

We perform the minimum-value calibration of an SIL assignment tool by first selecting the most pessimistic value from each of the categories that is used to describe the risk. For instance, a tool that uses consequence for one of its components might have a category with a range of PLL=0.1 to PLL=1.0. In this case, the most pessimistic value in the range is the PLL of 1.0. When all of the category parameters are selected in this fashion, the risk that is calculated for each box in the SIL assignment tool will be the maximum for the range. In the case of the risk graph, we must also select parameters that describe probability of avoidance and occupancy. These categories should be used to modify the consequence parameter so as to give an effective value for the maximum consequence parameter. For instance, suppose there is a consequence of one fatality where there is a 10 percent chance of occupancy and a 90 percent chance that the consequence might be avoided. This yields a PLL of $1.0 \times 0.1 \times 0.1 = 0.01$ as the effective maximum consequence parameter.

Once we have determined the maximum consequence and likelihood for a box in an SIL assignment tool, we can calculate the tolerable risk frequency for that event can using equation 10.3. Then using this tolerable frequency result, we can use equation 10.2 to calculate the PFD_{avg} of the SIF that is required to reduce risk to a tolerable level.

10.7.2 Prototypical Value Calibration

The prototypical value calibration process is similar to the minimum value calibration except that here one selects a single value to be typical of the entire range instead of selecting the maximum risk value. These single prototypical values are then used in the required risk reduction calculations to generate the SIL to be assigned to each box. For instance, a likelihood table might include three categories, as shown in table 10.10.

10.8 SIL Assignment Based on Environmental Consequence

So far in this chapter, we have focused on assigning SILs based solely on consequence as defined in terms of personnel injury and loss of life. As was described in section 6.1, however, a good risk management program should consider a broader range of consequences. These include business interruption, property damage, damage to sensitive environments, third-party liability, and other tangible or intangible losses that might result from the accidental release of chemicals or energy from a process.

Example 10.11

Problem: An organization is assembling a risk graph for use in its SIL selection process. The risk graph is shown at the end of this problem statement. The company's tolerable risk criteria include a maximum individual risk of fatality of 9.0×10^{-5} per year. The maximum societal risk is anchored at the individual risk-of-fatality criteria and uses a risk aversion factor of 2. The following ranges have been selected for some categories that describe the risk. The consequence category C_B has a range of PLL=0.3 to 1.5; the occupancy category F_A has a range of 0.0 to 0.1; the probability of avoidance category P_A has a range of 0.0 to 0.1, and the demand rate category W_2 has a range of 0.03 to 0.3 times per year. Based on this information, calibrate the box marked with an X in the following risk graph.

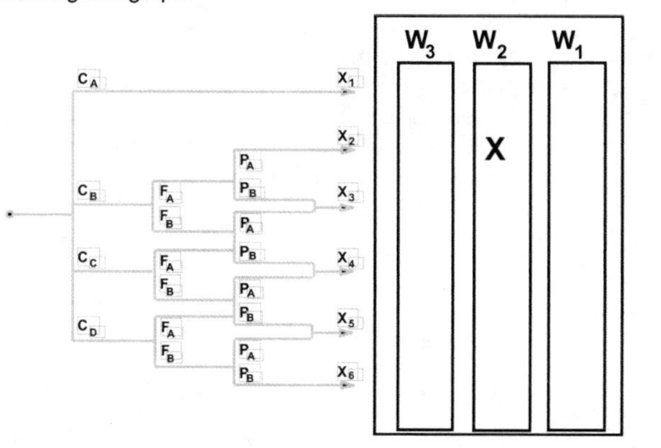

Solution: Based on the information provided in the statement of the problem, the maximum consequence would be the product of the maximum consequence in the C_B category range, the maximum occupancy in the F_A range, and the maximum probability of not avoiding the accident within the P_A range. As the following equation shows, the resulting effective maximum consequence in terms of probable loss of life, modified for avoidance and occupancy, is 0.015:

$$C_{MAX} = 1.5 \times 0.1 \times 0.1 = 0.015$$

If we enter into equation 10.3 the effective consequence we just calculated and the tolerable risk guidelines of the facility we gain a tolerable frequency for this event of 6.0×10^{-3} per year. Note that, as in example 10.10, because PLL<1 a risk aversion factor of 1 should be used.

$$f_{Tolerable} = \frac{f_{Tolerable,\ Individual}}{PLL^{\alpha}} = \frac{9.0 \times 10^{-5}}{0.015^{1}} = 6.0 \times 10^{-3}$$

The likelihood (or demand rate) is taken as the most frequent likelihood in the W_2 range, or 0.3 times per year. If we enter this likelihood value, together with the tolerable frequency we just calculated, into equation 10.2 we get the required PFDavg of the SIF, which in this case is 2.0×10^{-2}:

$$PFD_{SIF} = \frac{f_{Tolerable}}{f_{Event,\ No\ SIS}} = \frac{6.0 \times 10^{-3}}{0.3} = 2.0 \times 10^{-2}$$

A PFD of 2.0×10^{-2}, or RRF of 50, corresponds to an SIL of 2, as we discussed in section 10.5. Therefore, we should assign SIL 2 to the risk graph box that contains the "X."

Table 10.10 Typical Likelihood Categories: Prototypical

Category	Typical Frequency
Low	1 x 10^{-4} per year
Moderate	1 x 10^{-3} per year
High	1 x 10^{-2} per year

Example 10.12

Problem: An organization is assembling a hazard matrix for use in its SIL selection process, as shown at the end of this problem statement. The company's tolerable risk criteria include a maximum individual risk of fatality of 9.0 x 10^{-5} per year. The maximum societal risk is anchored at the individual risk-of-fatality criteria and uses a risk aversion factor of 2. A PLL of 1.5 is typical of the consequence we might expect for the "serious" category. Table 10.10 is used to select likelihood categories. What SIL should be placed in the box marked with an "X" in the following hazard matrix?

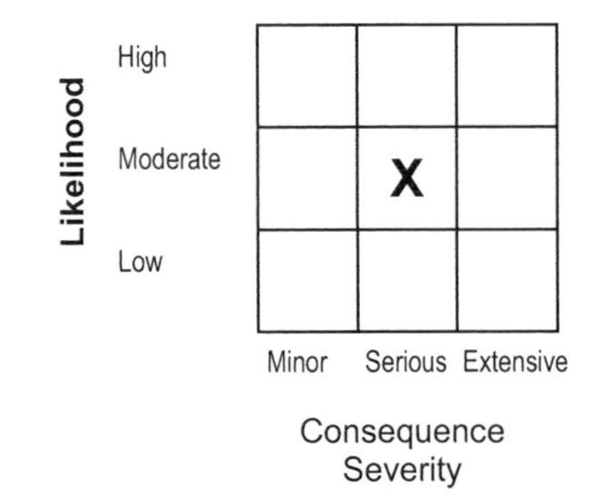

Solution: The prototypical consequence of the identified hazard matrix box is a PLL of 1.5. This consequence is then combined with the organization's tolerable risk criteria using equation 10.3 to yield a tolerable event frequency of 4.0x10^{-5} for that prototype "serious" consequence:

$$f_{Tolerable} = \frac{f_{Tolerable,\ Individual}}{PLL^{\alpha}} = \frac{9.0 \times 10^{-5}}{1.5^2} = 4.0 \times 10^{-5}$$

The likelihood that is typical of the "moderate" category is 1.0x10^{-3} times per year. Entering into equation 10.2 both the likelihood value and the tolerable frequency we just calculated, we compute the required PFDavg of the SIF, which in this case is 4.0x10^{-2}:

$$PFD_{SIF} = \frac{f_{Tolerable}}{f_{Event,\ No\ SIS}} = \frac{4.0 \times 10^{-5}}{1.0 \times 10^{-3}} = 0.04$$

A PFD of 0.04, or RRF of 25, corresponds to an SIL of 2. Therefore, we should assign SIL 2 to the hazard matrix box that contains the "X."

Typically, the consequences that might result from a chemical or energy release are separated into three categories: personnel safety, environmental impact, and financial losses. Personnel safety describes consequences that will cause injuries and fatalities to both onsite workers and offsite persons. The safety of personnel is usually the highest priority when risk reduction decisions are made. It is also the consequence that receives the most regulatory scrutiny. Thus, the previous parts of this chapter focused on assigning SILs based on personal injury and loss of life.

Environmental impact has increasingly become another category that deserves special attention. Because of the evolving regulatory and political climate in most countries, environmental damage must be given additional specific consideration. Damage to an environmentally sensitive area is often considered much more significant than simply the cleanup cost or removal of the spilled chemicals. In many cases, damage to sensitive environmental receptors cannot be reversed. As a result, the risk reduction decision-making process often gives environmental concerns a special priority that takes into account factors beyond the immediate financial cost of cleanup.

All other consequence types are usually combined into the category of financial loss. For these consequences, such as equipment damage and business interruption, it is a straightforward process to convert the loss into financial terms and one that is uncomplicated by political and regulatory issues. However, in many moderate-scale accidents, it can be very difficult to assess how severe the business interruption will be depending on which particular equipment was damaged and how critical it was to the overall production rates.

When using an SIF to decrease the likelihood or consequence of damage to the environment, there are two available options for analysis. The first is to include cleanup costs and fines from regulatory authorities as a cost and to include the environmental consequences and any other associated corporate image losses in the financial analysis. For the reasons stated earlier, some organizations do not use this approach, preferring the second option of assigning a separate integrity-level category to the SIF to reflect environmental concerns. This category is sometimes referred to as *Environmental Integrity Level,* or EIL. The EIL is assigned using the same tools as in SIL assignment (i.e., risk graph and hazard matrix), but the categories of consequence are described in terms of environmental loss. Table 10.11, which is based on information in the IEC 61511-3 standard, shows some typical consequence categories for an environmental loss risk graph.

The process for assigning an EIL is identical to the process for SIL assignment except that the category for consequence is selected from a different table, which is written in terms of environmental consequences. If the analyst is using a risk graph for the assignment, he or she should also consider very carefully whether to take any credit for low occupancy or probability of avoidance.

Table 10.11 Typical Environmental Consequence Categories: Risk Graph

Category	Description	Additional Comments
C_A	A release with minor damage that is not very severe but is large enough to be reported to plant management.	A moderate leak from a flange or valve, small-scale liquid spill, small-scale soil pollution without affecting groundwater.
C_B	A release within the plant fence line causing significant damage.	A cloud of noxious vapor traveling beyond the unit after a gasket blowout or compressor seal failure.
C_C	A release outside the plant fence line causing major damage that can be cleaned up quickly without significant lasting consequences.	A vapor or aerosol release, with or without liquid fallout, that causes temporary damage to plants or wildlife.
C_D	A release outside the plant fence line causing major damage that cannot be cleaned up quickly or has lasting consequences.	A liquid spill into a river or sea. A vapor or aerosol release, with or without liquid fallout, that causes lasting damage to plants or wildlife. Fallout of solids (dust, catalyst, soot, ash). A liquid release that could affect groundwater.

It is important to remember that you should not, technically speaking, refer to the integrity levels chosen for financial or environmental reasons as safety integrity levels or SILs. The term *SIL* has a specific regulatory and standards-based meaning that relates to personnel safety. If you choose a high integrity level for environmental or financial reasons and then list it as an SIL, the owner of the process may be forced to comply with additional safety regulations and face additional scrutiny from regulators. This is the case even though the risk that motivated the high integrity level selection was not related to personnel safety. We therefore recommend that if you determine multiple integrity levels based on safety, environmental, and financial concerns, you should list them all separately in the safety requirements specification. This process will establish and document the specific requirements for personnel safety, and separate them out from the requirements as a whole, which may be driven by environmental or financial considerations.

Example 10.13

Problem: A hazard assessment team considers a recommended safety instrumented function (SIF) whose purpose is to prevent the release of oil from a storage section of a process. The expected demand rate is 0.1 times per year without considering the SIF. If the release were to occur, a large amount of oil would be spilled, damaging the soil both on and off site. It is estimated that the spill would require a cleanup effort costing $3 million, but would not cause a lasting impact on the environment. Using the risk graph in figure 10.2 and table 10.11 for the consequence categories, what integrity level should be selected for environmental reasons?

Solution: Based on the problem's description, the consequence in question falls into the C_C category in table 10.11. The key parameters for this decision are that the spill would impact the environment off site, but can be cleaned up without lasting negative impacts. Considering the other parameters of the risk graph in this case, the environmental receptors are always present, so no credit for avoidance or lack of occupancy can be taken. As a result, the team chooses category F_B for occupancy and P_B for probability of avoidance. The demand rate of 0.1 times per year falls into the W_2 category. Following from the starting point and using the selected categories yields row X_5. Since W_2 is selected, the team chooses the second column. The selected box contains integrity level 3, as shown here:

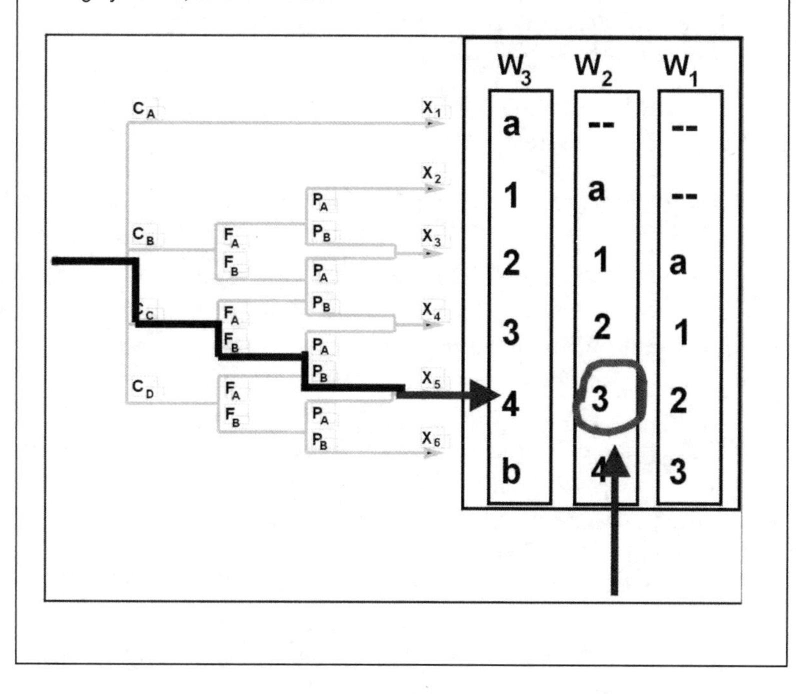

10.9 SIL Assignment Based on Financial Consequence

Reducing risk for the sake of minimizing financial losses is a very common application of SIFs since most accidents in the process industries have a significant financial component. Financial losses can be caused by a number of sources, a few of which are listed here:

- Damaged equipment that must be replaced
- Lost revenue caused by the inability to produce products (business interruption)
- Penalties imposed by customers for breaching supply contracts
- Third-party liability for injuries and fatalities resulting from process accidents
- Fines from regulatory agencies for personnel injuries and fatalities
- Cleanup costs for environmental damage resulting from spills and releases
- Fines from regulatory agencies due to environmental contamination
- Lost revenue because sales decreased as a result of public outrage at accident outcomes
- Expenses for public relations to repair damaged organizational image

While many of these items contribute to the total financial loss of an accident, most analysts focus on the sum of property damage and business interruption. The remaining sources of financial loss are only considered on a case-by-case basis, since the property damage and business interruption losses constitute the preponderance of the financial burden. In addition, many organizations are unwilling to include third-party liability for injuries and fatalities into their financial considerations, leaving them instead strictly to the SIL selection effort.

Like assigning SILs, assigning integrity levels based on financial concerns (sometimes referred to as *Financial Integrity Level*, or FIL) can be performed using either a qualitative or quantitative approach. The qualitative approach is to assign the degrees of financial loss to the consequence categories and then apply either the risk graph or hazard matrix approach to select the FIL. The quantitative approach is to calculate the costs associated with employing an instrumented function to reduce risk and then to compare those costs with the financial benefit of the instrumented function using cost-benefit analysis.

10.9.1 Category Based

The process for assigning an FIL is identical to the process for assigning an SIL, with one exception. The category for consequence is either selected from a different table, written in terms of financial consequences, or selected by reviewing the financial components of a single consequence table in which all of the considerations have been included in one place. For instance, consider table 10.3 in the context of a hazard matrix. Using this table to categorize financial loss would result in a category assignment of "serious" if the financial loss were near $1 million. If the analyst were using a risk graph to make the assignment, he or she should also consider very carefully whether any credit should be taken for low occupancy or probability of avoidance.

Example 10.14

Problem: A team of engineers and operators has analyzed an instrumented function that was recommended during a hazard analysis study. The objective of the function is to protect a compressor from serious damage. The compressor damage incident will occur at a frequency of about 1.3×10^{-3} per year if no instrumented function is present. The damage to the compressor as a result of this event will mean the compressor must be replaced at a cost of $6 million, but there will be no environmental or personnel impacts. Using tables 10.3 and 10.4 and figure 10.1, select the required FIL.

Solution: This problem asks for the FIL of an instrumented function whose consequence is $6 million in equipment damage, which is assigned a category of "extensive" based on table 10.3. The likelihood of the event falls into the "moderate" category according to table 10.4. Selecting the moderate likelihood row and then the extensive consequence column from the matrix yields an SIL 3 (FIL 3), as shown here:

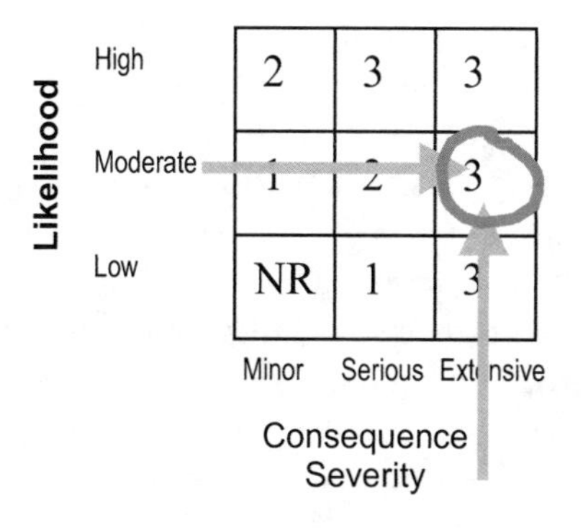

It is worth noting that this analysis did not consider the costs associated with the process downtime that would probably result, since a compressor of this size is unlikely to be immediately available. It is a relatively straightforward process to include these other financial considerations in the procedure for determining integrity level once they have been accurately estimated.

10.9.2 Cost-Benefit Analysis

Cost-benefit analysis is a well-respected and ubiquitous tool for financially analyzing the various options for performing a task. The topic of cost-benefit analysis can be quite complex and has been the object of substantial interest in the professional literature. The information we present in this section can therefore provide only a brief overview, targeted specifically toward the selection of integrity level. Readers interested in learning more about cost-benefit analysis should consult *Plant Design and Economics for Chemical Engineers* by Peters and Timmerhaus. The basic process for performing a cost-benefit analysis is as follows:

1. List the alternatives for solving a problem.
2. Calculate the costs for implementing each option.
3. Calculate the benefits to be obtained by employing each option.
4. Calculate the benefit-to-cost ratio or other comparable decision variable for each option.
5. Employ the option with the highest benefit-to-cost ratio or best decision-variable value.
6. If you are employing benefit-to-cost ratio, the benefit may be gained from employing multiple options. In this case, consider those options in order of decreasing benefit-to-cost ratio using a marginal benefit-to-cost ratio. This can be calculated by dividing the increase in benefit by the increase in cost for each relative to a reference alternative project option.

Many different cost-benefit measures or decision variables exist, including marginal rate of return, net present value, cash flow basis, risk-weighted rate of return, and so on. All of these can be applied to assessing the optimal option for reducing financial risk. They primarily differ in the way they assign the time value of money and other financial risk components. These finer details are beyond the scope of this text, but the key general concept of finding the most financially attractive option is important in justifying the value and use of safety instrumented systems.

For risk reduction projects, benefits are actually manifested as a decrease in the expected value of loss. For instance, a potential explosion might have a consequence of $10 million and an expected frequency of once in 10,000 years. In this case, the expected value of loss is $1,000 per year. A decrease in the event frequency to 10 percent of the original rate would lower the annual expected value of loss to $100 per year. The difference between the two—or $900 per year—would be the benefit derived from the 90 percent decrease in frequency.

The cost of implementing risk reduction with instrumented functions has two components: the installed cost of the equipment (including maintenance) and the cost of nuisance shutdowns. The cost of equipment should not only take into account the capital cost of purchasing the equipment, but also the cost of installation, maintenance, and periodic function tests. The overall cost should then be annualized to help make comparisons with the annualized benefits.

The second main component in the cost of an instrumented function is the cost of nuisance trips. The cost of nuisance trips results from the fact that business can be interrupted and off-spec product may be created that requires either re-work or disposal. You can calculate the nuisance trip cost on an annual basis by multiplying the average cost of a nuisance trip by the nuisance trip rate. The cost of the trip will include the lost production per unit time multiplied by the duration of the trip added to the fixed disposal or re-work costs per episode.

Once you know the costs and benefits of each option, you can calculate the benefit-to-cost ratio of each option by dividing the annualized benefits by the annualized costs. The resulting ratio describes the cost efficiency of the project. Any projects that result in a benefit-to-cost ratio greater than one are said to be cost-effective because they are returning more than one dollar in benefits for each dollar invested to implement the solution. Those options that have a benefit-to-cost ratio of less than one are not cost-effective.

It is likely that many of the options listed in the cost-benefit analysis will have a benefit-to-cost ratio that is greater than one. In this case, the organization should implement the option with the highest benefit-to-cost ratio first, as that is the option with the highest return on investment. Other options on the list that have a benefit-to-cost ratio greater than one may also be effective in reducing risk, but that is not always the case. For the rest of the items on the list with a benefit-to-cost ratio greater than one, the organization needs to do an incremental analysis of costs and benefits. Using the incremental or marginal analysis, you would weigh the increased cost of applying additional options against the additional benefits relative to a reference alternate project option. If the incremental or marginal ratio of benefit to cost is still greater than one, the organization should employ that option.

It should be noted that cost-benefit analysis does not result in the selection of an integrity level category. Instead, it results in the selection of the most cost-effective equipment configuration.

10.10 Selecting from Multiple Integrity Level Categories

When selecting an integrity level that is based on multiple categories of consequence, such as SIL, EIL, and FIL, you must also select an overall integrity level to guide the equipment design. You will usually make this assignment of overall integrity level by selecting the highest of the individual integrity levels.

Example 10.15

Problem: A team of engineers is considering an instrumented function to prevent an explosion in a manufacturing cell. While this explosion will not have any personnel impact, it will cause $2.5 million in property damage at a frequency of once in 200 years if no instrumented protective function is used. Three instrumented system architectures are being considered, as shown in the following table. Which alternative should the team select on a cost-benefit basis?

Option	Pressure Sensors	Logic Solver	Valves	RRF	Annual Cost
1	1oo1	Relay	1oo2	55	$2,400
2	2oo3	Relay	1oo2	450	$5,430
3	2oo3	PES	1oo2	1100	$9,786

Solution: The annualized costs for each option are already provided in the table. The annualized benefits for each option can be calculated by determining the decrease in expected value for each option. After this is done, the team can calculate benefit-to-cost ratio.

The annual expected value (AEV) of the property damage is the same in all three cases since a single outcome is being considered. This AEV before any protective function is then:

$$AEV = \frac{\$2,500,000}{200} = \$12,500$$

The AEV for each protective function is then calculated by dividing the initial AEV by the risk reduction factor for that protective function. This new AEV is then subtracted from the initial value to give the annual benefit of each option. Finally, the ratio is taken to arrive at the decision variable of interest in the right-hand column of the following spreadsheet:

	Annual Exp Value	RRF	AEV After RRF	Annual Benefit	Annual Cost	Ben-Cost Ratio
1	$12,500	55	$227	$12,273	$2,400	5.11
2	$12,500	450	$28	$12,472	$5,430	2.30
3	$12,500	1100	$11	$12,489	$9,786	1.28

Based on the results of the analysis, the system that should most likely be incorporated is the first system, which includes 1oo1 (one-out-of-one) pressure sensors, relay-based logic, and 1oo2 voting valves. The analysis also shows that options 2 and 3 also had benefit-to-cost ratios greater than one, so the team should also explore these options further with incremental analysis. Taking option 1 as the reference alternate project option for comparison, option 2 gives an incremental or marginal cost increase of $5,430 – $2,400, or $3,030. The increased or marginal benefit between options 1 and 2 is $12,472 – $12,273, or $199. Therefore, the incremental or marginal benefit-to-cost ratio for moving from option 1 to option 2 is $199 / $3,030, or 0.07. Since the marginal benefit-to-cost ratio is less than one, upgrading from 1oo1 pressure sensors in option 1 to 2oo3 pressure sensors in option 2 is not cost-effective.

This same type of analysis will show that the move from a relay-based system to a programmable electronic system (PES) is also not cost-effective. In this case, after performing this full evaluation, the team should recommend that option 1 be employed.

Example 10.16

Problem: A process hazard assessment team has selected integrity levels for a proposed instrumented function in terms of personnel safety, environmental impact, and financial considerations. The results of the process yielded: SIL=1, EIL=2, and FIL=3. What overall integrity level should the organization select for the design of the instrumented function?

Solution: The integrity level of the instrumented function that is required to meet all of the analyzed goals would be 3. The integrity level of 3 will meet the risk reduction for not only financial but also environmental and safety considerations. In addition, because the financial criteria determined the design in this case, the integrity level of 3 actually provides a net financial benefit relative to the other two options. It is important to note that the team should present all three of the individual integrity level figures in the safety requirements specification in addition to the overall integrity level. Doing so will document the basis for selecting the integrity level and demonstrate how that integrity level is actually related to personnel safety.

While selecting the highest of the individual integrity levels works well in most cases, there are situations where it may be inappropriate. The overall risk of a process is not about selecting among categories of risk to determine the single highest one; it is a summation of the risks in all of the different areas. In example 10.16, the integrity level categories were all different, and thus the safety and environmental components of the risk were small in comparison with the financial aspects. What if the SIL were 3, the EIL were 2, and the FIL were also 3? In this situation, it is possible that when safety, environmental, and financial considerations are summed together, the overall integrity requirement might be equivalent to SIL 4.

While few organizations perform this type of analysis, it is becoming more prevalent as analysts become more comfortable with risk measures, and as the benefits of better risk management practices become apparent. Summing different risk types is done by using a process called *multi-attribute utility*. In general, losses of all types represent a loss of utility, or value. Since different types of losses have different attributes, all loss types need to be converted into a uniform basis for the purpose of comparison and summation. Typically, this uniform basis is financial, but there are certain challenges in translating personnel safety into monetary terms, as we discussed in chapter 3. Once you know the uniform basis loss, you can categorize it and incorporate it into a risk graph or hazard matrix. Alternatively, you can use the uniform financial loss directly to perform a cost-benefit analysis for the purpose of selecting the overall most cost-effective equipment architecture for the instrumented function.

10.11 Summary

This chapter presented a variety of methods for assigning safety integrity levels (SILs) to safety instrumented functions (SIFs) by using the level of risk inherent in the process relative to the amount of risk tolerable to the process owner. The selection methods fall broadly into two categories, qualitative and quantitative. Qualitative methods describe risk in terms of broad categories of the risk's consequence and likelihood components. The qualitative methods described in this chapter included hazard matrix and risk graph. Hazard matrices describe the risk of a specific harmful outcome of the process in terms of consequence and likelihood categories. These categories are converted into an SIL by a two-dimensional matrix. Risk graphs also consider the likelihood and consequence components, but they consider other, more detailed factors that contribute to these components, such as occupancy and probability of escape. In both of the methods we described, the tolerable level of risk is not explicitly stated, but rather implied in the structure of the tool.

Quantitative methods for assigning SILs are also widely used. These methods require that the tolerable risk level be stated explicitly as a numerical target. This target will either be related to a category of consequence or to other risk criteria such as individual risk of fatality and societal risk. Once you know the frequency target, you can calculate the required risk reduction (in terms of probability of failure on demand of the SIS) using probability multiplication. The SIL selected through this method should be one SIL number greater than the SIL that corresponds to the range of the calculated risk reduction factor. The exception to this is when the SIL category is specifically combined with a required risk reduction factor as part of the safety requirements specification. When the more detailed specification of both aspects is not provided, the higher SIL number is needed to ensure that all systems that meet the specified SIL have achieved the required amount of risk reduction.

In addition to personnel safety, instrumented functions are used to protect the environment and to guard against other consequences that can impact the financial well-being of an organization. For these consequence categories, integrity levels are often selected in addition to the safety integrity level targets. For example, environmental impact integrity levels are often selected by using the same qualitative tools of risk graph and hazard matrix, but different tables are used to categorize consequence. You can also select integrity levels based on financial considerations by using qualitative categorization methods. However, the most accurate results are produced when you use cost-benefit analysis to directly compare different specific equipment configuration options.

10.12 Exercises

10.1 If a risk reduction of 50 is required to move the risk of a process into the tolerable range, what SIL should be selected as a single specification?

10.2 An SIF is used in the continuous mode of operation and has a failure rate of 5.0×10^{-6} per hour. What is the SIL of this SIF with regard only to failure rate?

10.3 Referring to tables 10.3 and 10.4 and figure 10.1, make SIL assignments using the hazard matrix process for the following situations:

a. Consequence is multiple fatalities (~7), and likelihood is remote (~10^{-6} per year).

b. Consequence is minor injury, and the likelihood is remote (~10^{-6} per year).

c. Consequence is expected to be a single fatality that occurs with a likelihood of once in 25 years without considering the SIF.

10.4 Using tables 10.3, 10.5, 10.6, 10.7, and 10.8 and figure 10.2, make SIL assignments for the following situations using the risk graph method:

a. Consequence is multiple fatalities (~7), and likelihood is remote (~10^{-6} per year). There is no possibility of avoidance, but the area is not normally occupied.

b. Consequence is minor injury, and the likelihood is remote (~10^{-6} per year). There is a good possibility of avoidance as the operator is alerted to hazardous conditions and has ample time to leave the area. The area is normally occupied.

c. Consequence is expected to be a single fatality that occurs with a likelihood of once in 25 years without considering the SIF.

d. Consequence has a probable loss of life of 0.98 and a demand rate of 0.45 times per year.

e. Consequence is $7 million in business interruption losses, negligible equipment damage, and no injury or fatality. This consequence will occur with a likelihood of once in 700 years without considering the effect of the proposed SIF.

10.5 Referring to table 10.9, make SIL assignments for the following situations using the frequency-based assignment method. Assign the SIL both in terms of SIL only and SIL with RRF:

a. Consequence is "extensive," and the likelihood is 3.4×10^{-3} per year.

b. Consequence is "minor," and the likelihood is 9.7×10^{-7} per year.

c. Consequence is "serious," and the likelihood is 8.0×10^{-2} per year.

10.6 An organization has a maximum individual risk-of-fatality criterion of 2.5×10^{-5} per year and uses a risk-neutral approach to societal risk (i.e., risk aversion factor of 1). A team at this organization has performed a QRA (quantitative risk assessment) to determine the risk posed by a process for which an SIF is being considered to reduce risk. The event frequency without the SIF was determined, using LOPA, to be once in 180 years. The consequence of the event was determined, using quantitative consequence models, to be a probable loss of life of 2.4 persons. What SIL should be selected for the SIF under consideration?

10.7 An organization has a maximum individual risk-of-fatality criterion of 1.0×10^{-5} per year, and it uses a risk-neutral approach to societal risk (i.e., risk aversion factor of 1). The organization wants to use the risk graph shown in the following figure to select its SIL, and has defined its categories as shown in the figure. Calibrate the risk graph for the appropriate SIL values in each risk box.

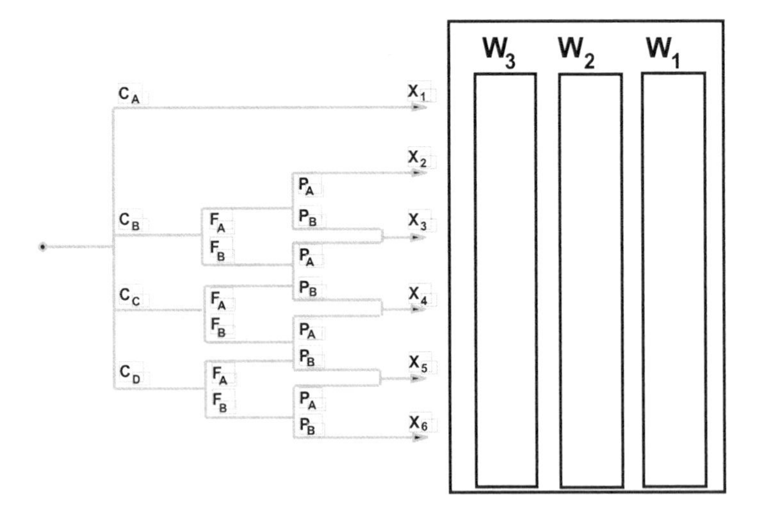

Consequence Categories

Category	Description
C_A	PLL = 0.001 to 0.01
C_B	PLL = 0.01 to 0.1
C_C	PLL = 0.1 to 1.0
C_D	PLL = 1.0 to 10

Occupancy Categories

Category	Description
F_A	Occupied less than 10% of time
F_B	Continuously occupied

Probability of Avoidance Categories

Category	Description
P_A	Can be avoided 90% of time
P_B	No chance of avoidance

Demand Rate Categories

Category	Description
W_1	0.01 to 0.1 per year
W_2	0.1 to 1 per year
W_3	1 to 10 per year

10.8 Reverse engineer the calibration on the following hazard matrix to determine the underlying maximum individual risk of fatality and risk aversion factor.

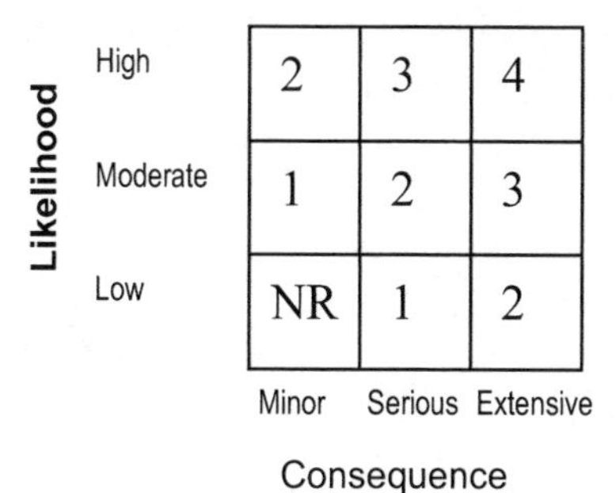

Consequence Categories

Category	Description
Minor	PLL = 0.001 to 0.01
Serious	PLL = 0.01 to 0.1
Extensive	PLL = 0.1 to 1.0

Likelihood Categories

Category	Description
Low	0.001 to 0.01 per year
Moderate	0.01 to 0.1 per year
High	0.1 to 1 per year

10.9 Considering tables 10.6, 10.7, 10.8, and 10.11 and figure 10.2, make EIL assignments for the following situations using the risk graph assignment method:

a. Consequence is a small release with little to no damage that is contained within the fence line, and the likelihood is once in 10 years.

b. Consequence is a large release that escapes the boundary of the plant. The impact of the accident will have a permanent effect on the surrounding environment. The likelihood of the accident without an SIF is once in 50 years.

c. A particular process upset can result in the release of 10,000 gallons of boiling water into a stream. If the release were to occur during the mating season of a certain endangered fish, it would cause the extinction of this breed of fish. The fish are only present near the plant during the mating season, which lasts two weeks per year. The likelihood of the particular process upset is once in 100 years.

10.10 List at least five different sources of financial loss.

10.11 A team selects an integrity level in terms of safety, environmental, and financial concerns with the following results: SIL=4, EIL=1, and FIL=3. What overall integrity level should the team select for this instrumented function?

10.12 When multiple types of integrity level are selected for an instrumented function, which should be listed in the safety requirements specification? Explain your answer.

10.13 An accident is likely to occur once in 50 years and has an expected loss of $55 million for the event. What is the greatest annual benefit that can be expected from an integrity level 1 system? The least?

10.14 What is the annual cost of the following SIF?

Installed cost of equipment	$18,600
Cost per test	$780
Test interval	3 months
Annual maintenance	$350
Nuisance trip rate	1/25 per year
Cost per nuisance trip	$30,000 for contaminated batch disposal
Nuisance trip downtime	16 hours
Profit from production	$450,000 per day

10.13 References

1. ANSI/ISA-84.01-1996 - *Application of Safety Instrumented Systems for the Process Industries.* Research Triangle Park, NC: ISA, 1996.
2. Center for Chemical Process Safety. *Guidelines for Safe Automation of Chemical Processes.* New York: American Institute of Chemical Engineers, 1993.
3. International Electrotechnical Commission. *Functional Safety of Electrical / Electronic / Programmable Electronic Safety-Related Systems,* IEC 61508. Geneva: International Electrotechnical Commission, 1998.
4. International Electrotechnical Commission. *Functional Safety: Safety Instrumented Systems for the Process Industry Sector,* Draft Standard 61511 (CDV 12/6/00). Geneva: International Electrotechnical Commission, 2000.
5. Peters, M. S., and K. D. Timmerhaus. *Plant Design and Economics for Chemical Engineers.* New York: McGraw-Hill, 1991.

APPENDIX A

Derivation of Equations

This appendix contains derivations of some of the equations in the text to provide additional understanding of the mathematical basis of the concepts presented in the earlier sections.

A.1 Derivation—SIL Assignment Equation

The derivation of the SIL assignment equation (equation 10.2) begins with the law of probability multiplication, as shown in equation A.1, which is true by definition for two independent events from basic probability mathematics.

$$P_{(AandB)} = P_A P_B \tag{A.1}$$

This is then applied to the problem of SIL assignment. Here one starts with the probability of an unwanted event taking place without any SIS present, defined as $P_{Unmitigated\ Event}$. This is combined with the definition of a successful SIS action (preventing the unwanted event from occurring) and the failure of that action (allowing the unwanted event to happen). Thus, in order for the unwanted event to occur with the SIS in place, $P_{Unmitigated\ Event}$ must be true AND $P_{SIS\ Failure}$ must also be true. The two independent probabilities are multiplied to give the $P_{Unwanted\ Event}$ as shown in equation A.2.

$$P_{UnwantedEvent} = P_{UnmitigatedEvent} P_{SISFailure} \tag{A.2}$$

For the purposes of the SIL selection problem, the terms whose probabilities are being discussed should be clarified. In order to make this problem useful, the probability at which an unwanted event is allowed to occur must be limited to a tolerable level. As such, the term $P_{UnwantedEvent}$ should be redefined as $P_{Tolerable}$. It is important to note that the tolerable probability discussed here refers to the probability at which the specific unwanted event is tolerable, as opposed to more general tolerable risk goals such as individual risk of fatality.

As defined in ANSI/ISA-84.01-1996 and IEC 61508/61511, the probability that an SIS will fail is defined as the probability of failure on demand (PFD). It is important to note that the PFD applies to an individual loop, called a safety instrumented function, as opposed to the entire system that includes all of the safety functions performed by a particular logic solver. This more precise definition of $P_{SIS\ Failure}$ is thus PFD_{SIF}. After making these substitutions, the relation can be expressed in the form shown in equation A.3.

$$P_{Tolerable} = P_{UnmitigatedEvent} PFD_{SIF} \tag{A.3}$$

Solving the equation for probability of failure on demand of the safety instrumented function yields equation A.4.

$$PFD_{SIF} = \frac{P_{Tolerable}}{P_{UnmitigatedEvent}} \tag{A.4}$$

The probability that an event will occur in the future is assumed to be a function of the frequency at which that event has historically occurred and the time interval over which the probability is calculated. The fundamental equation relating probability, frequency, and reference time interval is shown in equation A.5 for a constant historic event frequency. This assumption of constant frequency is consistent with general assumptions used in reliability engineering and is consistent with empirical evidence.

$$P = 1 - e^{-f \times t} \tag{A.5}$$

where:

f is the frequency of the event

t is the interval over which the probability is calculated

Using equation A.5 to describe both the probability of a tolerable event and the probability of the occurrence of an unmitigated event in equation A.4 yields equation A.6.

$$PFD_{SIF} = \frac{1 - e^{-f_{Tolerable} \times t_{Tolerable}}}{1 - e^{-f_{Unmitigated\ Event} \times t_{Unmitigated\ Event}}} \tag{A.6}$$

Taking the first two terms of a Taylor series expansion for each of the exponents yields equations A.7 and A.8. It is important to note that this is only valid for relatively small values of the f-t products in the exponentials.

$$e^{-f_{Tolerable} \times t_{Tolerable}} = 1 + (-f_{Tolerable} \times t_{Tolerable}) \tag{A.7}$$

$$e^{-f_{Unmitigated\ Event} \times t_{Unmitigated\ Event}} = 1 + (-f_{Unmitigated\ Event} \times t_{Unmitigated\ Event}) \tag{A.8}$$

Making the corresponding substitutions into equation A.6 yields:

$$PFD_{SIF} = \frac{1-(1+-f_{Tolerable} \times t_{Tolerable})}{1-(1+-f_{Unmitigated\ Event} \times t_{Unmitigated\ Event})} = \frac{f_{Tolerable} \times t_{Tolerable}}{f_{Unmitigated\ Event} \times t_{Unmitigated\ Event}} \quad (A.9)$$

The variable t represents the amount of time over which a probability is calculated. In all cases, solution of the problem requires that the probability for the tolerable event occurrence and the probability for the unmitigated event occurrence be evaluated over the same reference time interval. Therefore, $t_{Tolerable}$ is equal to $t_{UnmitigatedEvent}$. Since the two terms are divided in equation A.9, they will both drop out of the equation, yielding the final result in equation A.10.

$$PFD_{SIF} = \frac{f_{Tolerable}}{f_{UnmitigatedEvent}} \quad (A.10)$$

A.2 Derivation—Tolerable Event Frequency

The derivation of the equation for the tolerable frequency of an event based on that event's consequence and the maximum individual risk of fatality criteria begins with the mathematical definition of the tolerable societal risk curve.

A typical F-N curve is shown in figure 3.3. The curve is drawn on a log-log chart, with the number of fatalities on the x-axis and the cumulative event frequency on the y-axis. The limits of tolerance can be shown on the same plot as straight lines with a negative slope. Usually, two lines are drawn, with the upper line representing the *de manifestus* risk level (maximum limit for tolerance) and the lower line representing the *de minimus* risk level (below which risk is negligible). The area between these two lines is the ALARP region (See section 3.1 for a discussion of ALARP). In some cases, only a single line for *de manifestus* risk is shown. Figure A.1 shows a typical tolerable societal risk *de manifestus* curve. The curve is drawn on a linear chart using the log (fatalities) and the log (frequency) as the independent (x) and dependent (y) variables instead of a log-log chart to facilitate the derivation.

An equation that represents the curve shown in figure A.1 begins with one of the equations for a straight line in linear Cartesian coordinates, as shown in equation A.11.

$$y = mx + b \quad (A.11)$$

In equation A.11, y represents the log of the frequency (F) of an event and x represents the log of the number of fatalities (N) that would result from the event. The slope of the line is m and b is the y-intercept or the

Figure A.1 Typical Tolerable Societal Risk De Manifestus Curve

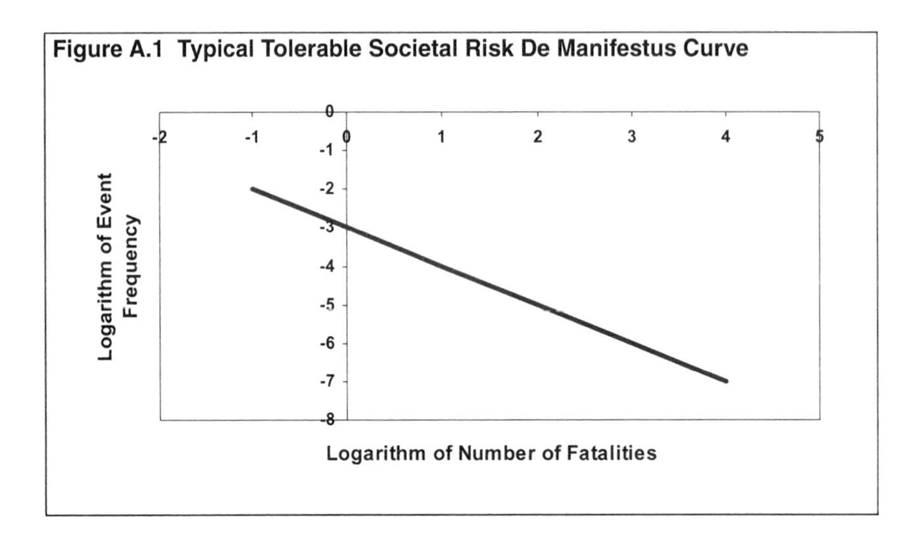

point where the line crosses the y-axis. Substituting these specific values into equation A.11 yields equation A.12.

$$\log(F) = m \times \log(N) + b \tag{A.12}$$

The frequency of interest is the tolerable frequency of an event with a specified number of fatalities. Therefore, the generic F is changed to the more specific F_{Tol}. The number of fatalities expected for an incident is called the probable loss of life (PLL), as described in section 3.4.

The slope of the tolerable societal risk line represents how heavily multiple fatality incidents are weighed against multiple single fatality incidents that would yield the same harm. This is customarily referred to as the risk aversion index and is shown with the Greek symbol "α." The slope of the tolerable societal risk curve is always negative, as the tolerability of an event will always decrease as the number of fatalities increases (e.g., a 100 fatality event will never be more tolerable than a 1 fatality event). However, risk aversion is customarily described as a positive number. Therefore, the slope is the negative of the risk aversion factor.

The y-intercept (b) is the point where the line crosses the y-axis. At this point, the value of x is zero and the value of N, or number of fatalities, is one. The maximum frequency at which a single fatality is tolerable is typically set in organizational risk criteria as the individual risk of fatality, and is represented as F_{Ind}. Substituting these values into A.12 gives A.13.

$$\log(F_{Tol}) = -\alpha \times \log(PLL) + \log(F_{Ind}) \tag{A.13}$$

Equation A.13 can then be rearranged to bring the risk aversion factor into the exponent of PLL, as shown in equation A.14.

$$\log(F_{Tol}) = \log(PLL^{-\alpha}) + \log(F_{Ind}) \tag{A.14}$$

Next, both sides of the equation are raised as the power of ten.

$$10^{(\log(F_{Tol}))} = 10^{(\log(PLL^{-\alpha})+\log(F_{Ind}))} \tag{A.15}$$

Then, the addition in the exponent on the right side of the equation is broken out.

$$10^{(\log(F_{Tol}))} = 10^{(\log(PLL^{-\alpha}))} \times 10^{(\log(F_{Ind}))} \tag{A.16}$$

The base and exponents of all of the terms of the equation are then simplified.

$$F_{Tol} = PLL^{-\alpha} \times F_{Ind} \tag{A.17}$$

Finally, the equation is rearranged to facilitate the use of a positive risk aversion factor, yielding the final equation.

$$F_{Tol} = \frac{F_{Ind}}{PLL^{\alpha}} \tag{3.2}$$

A.3 Derivation—Component Average Probability of Failure (Single Mode)

As discussed in section 5.5.4, the probability of failure on demand of a device depends on the frequency of testing. An untested device's PFD gets larger as time increases. For a constant failure rate, the relationship between failure rate and test interval is exponential, and is described by equation 5.11.

$$PFD_{Max} = 1 - e^{-\lambda \times t} \tag{5.11}$$

where:

PFD_{Max} is the maximum failure probability, which occurs at the end of the time interval

λ is the failure rate

t is the mission time, or time between complete function tests

The average failure probability over a time interval is calculated by integrating the failure probability function, as shown in equation 5.11, over the test interval T, and then dividing by the test interval, as shown in equation 5.12.

$$PFD_{avg} = \frac{\int_{t-0}^{t} (1 - e^{-\lambda \times t})dt}{T} \tag{5.12}$$

The exponential function in equation 5.12 can be written with a Taylor series expansion as follows:

$$e^{-\lambda \times t} = 1 + (-\lambda t) + \frac{(-\lambda t)^2}{2!} + \frac{(-\lambda t)^3}{3!} \ldots \tag{A.18}$$

When the value of $-\lambda t$ is small, the Taylor series can be truncated at the second term and substituted back into equation 5.12 to give:

$$PFD_{avg} = \frac{\int_{t=0}^{T} (1 - (1 - \lambda t))dt}{T} = \frac{\int_{t=0}^{T} (\lambda t)dt}{T} \tag{A.19}$$

This equation is then integrated to give equation 5.13 expressed using the time interval t.

$$PFD_{avg} = \frac{\lambda t}{2} \tag{5.13}$$

Acronyms

1oo1, 2oo3, etc.,	One out of one, two out of three, etc. control logic architecture
AIHA	American Industrial Hygiene Association
ALARP	As Low As Reasonably Practicable
BLEVE	Boiling Liquid Expanding Vapor Explosion
BPCS	Basic Process Control System
CPQRA	(Guidelines for) Chemical Process Quantitative Risk Analysis
E/E/PE	Electrical / Electronic / Programmable Electronic
EIL	Environmental Integrity Level
EPA	Environmental Protection Agency
ERPG	Emergency Response Planning Guidelines
ERRF	External Risk Reduction Facilities
EUC	Equipment Under Control
EV	Expected Value (of Loss)
FIL	Financial Integrity Level
HSE	(UK) Health and Safety Executive
IDLH	Immediately Dangerous to Life and Health
IEC	International Electrotechnical Commission
ISA	ISA—The Instrumentation, Systems, and Automation Society
LOPA	Layer of Protection Analysis
NIOSH	(US) National Institute of Occupational Safety and Health
OCA	Offsite Consequence Analysis
OSHA	Occupational Safety and Health Administration
P&ID	Piping and Instrumentation Diagram
PES	Programmable Electronic System

PFD	Probability of Failure on Demand or Process Flow Diagram
PSM	Process Safety Management
RMP	Risk Management Plan
RRF	Risk Reduction Factor
SIF	Safety Instrumented Function
SIL	Safety Integrity Level
SIS	Safety Instrumented System
SLC	Safety Life Cycle
SRS	Safety Requirements Specification

Glossary

The terms that are used to describe safety instrumented system engineering are very important. Unfortunately, there are not uniform definitions for these terms, as people who perform different tasks use the same terms to mean different things. This comes from the difficult overlap of the electrical and controls engineering fields with the actuarial and insurance loss prevention fields. Sometimes both definitions will be shown here to highlight the differences. The terms and definitions presented here are mainly derived from IEC 61508 and IEC draft 61511 to describe the SIS aspects and from CPQRA to define the generic loss prevention tasks.

1oo1, 2oo3 Control Architecture – a specific configuration of software or hardware elements within a system. The first number designates how many elements are required to signal a shutdown condition. The second number designates how many total elements are used in the system. This is also called a voting system.

ALARP (As Low As Reasonably Practicable) – the philosophy of dealing with risks that fall between an upper and lower extreme. The upper extreme is where the risk is so great that it is refused altogether, while the lower extreme is where the risk is, or has been made, so small as to be insignificant. This philosophy considers both the costs and benefits of risk reduction to make the risk "as low as reasonably practicable." (Source: IEC draft 61511.)

Basic Process Control System – system that responds to input signals from the process, its associated equipment, other programmable systems and/or an operator and generates output signals causing the process and its associated equipment to operate in the desired manner, but that does not perform any safety instrumented functions with a claimed SIL 1. (Source: IEC draft 61511.)

BLEVE (Boiling Liquid Expanding Vapor Explosion) – a specific type of fireball. A BLEVE can occur as the result of the situation where a vessel containing a pressurized liquid comes in direct contact with external flame. As the liquid inside the vessel absorbs the heat of the external fire, the liquid begins to boil, increasing the pressure inside the vessel to the set pressure of the relief valve(s). The heat of the external fire will also be directed to portions of the vessel where the interior wall is not "wet" with the process liquid. Since the process liquid is not present to carry heat away from the vessel wall, the temperature in this region (usually near the interface of the boiling liquid) will rise dramatically, causing the vessel wall to overheat and become weak. A short time after the vessel wall begins to overheat, the vessel can lose its structural integrity and a rupture will occur. After vessel rupture, a fireball will usually result with the external fire available as the ignition source.

Calibration – the method by which SILs are assigned to a risk graph or hazard matrix based on the level of risk that is tolerable to the owner of the process in question.

Consequence – the magnitude of harm or measure of the resulting outcome of a harmful event. One of the two components used to define a risk.

Consequence Analysis – the detailed procedure for estimating the magnitude of harm or measure of the resulting harmful event outcome. Depending on the nature of the consequence, especially those involving toxic, flammable, or explosive material releases, the calculation procedures can be extremely complicated.

Continuous Mode (High Demand Mode) – when demands to activate an SIF are frequent compared to the test interval of the SIF. The process-specific IEC draft 61511 standard defines this mode when the demands to activate the SIF are more than two times per proof test interval. The more general IEC 61508 also includes systems where the demands to activate are more than once per year as part of this mode. The continuous mode of operation is common in the machine industry and avionics. In the continuous mode, the frequency of an unwanted accident is essentially determined by the frequency of an SIF failure. When the SIF fails, the demand for its action will occur in a much shorter time frame than the function test, so speaking of its failure probability is somewhat misleading. Essentially all of the dangerous faults of an SIF in continuous mode service will be revealed by a process demand instead of a function test. (Source: IEC 61508 and draft 61511.)

Demand Mode (Low Demand Mode) – when demands to activate the SIF are infrequent compared to the test interval of the SIF. The process-specific IEC draft 61511 standard defines this mode for when the demands to activate the SIF are less than two times per proof test interval. The more general IEC 61508 also includes systems where the demands to activate are less than once per year as part of this mode. The demand mode of operation is the most common mode in the process industries and the only mode of operation considered in the ANSI/ISA-

84.01-1996 standard. When defining the SIL for the low demand mode, an SIF's performance is measured in terms of average probability of failure on demand (PFD_{avg}). In this demand mode, the frequency of the initiating event, modified by the SIF's probability of failure on demand, determines the frequency of unwanted accidents. (Source: IEC 61508 and draft 61511.)

De Manifestus Level – the maximum level of tolerable risk in the context of the ALARP principle. Above this level the risk is unacceptable and must be mitigated or avoided under all circumstances. Below this level, the risk must be reduced only as it is reasonably practical to do so.

De Minimus Level – the minimum level of risk in the context of the ALARP principle where active risk reduction must be considered. Above this level the risk must be reduced where it is reasonably practical to do so. Below this level the risk is generally acceptable and no risk reduction is required.

Effect Zone – the physical area in which a harmful effect is felt by a receptor. For a toxic release, the area over which the airborne concentration exceeds some level of concern. For a physical energy release, the area over which a specified overpressure criterion is exceeded. For thermal radiation effects, the area over which an effect is based on a specified damage criterion [e.g., a circular effect zone surrounding a pool fire resulting from a flammable liquid spill, whose boundary is defined by the radial distance at which the radiative flux from the pool fire has decreased by 5 kW/m^2 (approximately 1600 $BTU/hr\text{-}ft^2$)]. (Source: CPQRA.)

Electrical/Electronic/Programmable Electronic System (E/E/PES) – system for control, protection, or monitoring based on one or more programmable electronic devices, including all elements of the system such as power supplies, sensors, and other input devices, data highways and other communication paths, and actuators and other output devices. (Source: IEC 61508.)

Environmental Integrity Level (EIL) – when SIFs are used to decrease the likelihood or consequence of damage to the environment, a separate integrity level category can be assigned to the SIF as a result of environmental concerns. This Environmental Integrity Level or EIL assignment is made using the same tools as SIL assignment (i.e., risk graph and hazard matrix), but the categories of consequence are described in terms of environmental loss.

Events (Complementary) – a pair of complementary events defines a situation where only one of a set of two outcomes is possible. If one event occurs, then by definition the other event cannot. For instance, when a rounded edge coin is tossed, the result must be either heads or tails. There is no way that the toss can result in both heads and tails, or neither. Therefore, heads and tails form a pair of complementary events.

Events (Independent) – events with outcomes that are not related in any way. When events are independent, the outcome of a prior event does not affect the probability of another subsequent event. For example, the fact that one coin toss resulted in heads does not impact the outcome of the next coin toss. Event independence is a requirement for probability multiplication.

Event (Initiating) – the first event in an event sequence (e.g., the stress corrosion resulting in leak/rupture of the connecting pipeline to the ammonia tank). (Source: CPQRA.)

Event (Intermediate) – an event that propagates or mitigates the initiating event during an event sequence (e.g., improper operator action fails to stop the initial ammonia leak and causes propagation of the intermediate event to an incident; in this case the intermediate event outcome is a toxic release). (Source: CPQRA.)

Events (Mutually Exclusive) – a set of events where only one of a set of outcomes is possible. No two mutually exclusive outcomes can occur simultaneously. For example, when one die is tossed the outcomes (1, 2, 3, 4, 5, or 6) are mutually exclusive. The roll of the die cannot result in 2 and 3 at the same time. Complementary events are a subset of mutually exclusive events where there are only two possible outcomes in the set.

Event Tree Analysis – a particular method of fault propagation modeling. The analysis constructs a tree-shaped graphical representation of the chains of events leading from an initiating event to various potential outcomes. The tree expands from the initiating event in branches of intermediate propagating events. Each branch represents a situation where a different outcome is possible. After including all of the appropriate branches, the event tree ends with multiple possible outcomes.

Expected Value – a form of risk integral where the consequences of the unwanted events are expressed on a uniform financial basis. This form is often used in insurance loss prediction calculations.

Explosion (Physical) – the result of sudden catastrophic rupture of a high-pressure vessel. The blast wave is caused when the potential energy stored in the high-pressure vessel is transferred to kinetic energy when that material is released. The effect zone is determined by the quantity of energy released and the blast shock wave overpressure resulting from the explosion.

Explosion (Vapor Cloud) – the result of ignition of a cloud of flammable vapor, when the flame velocity is high enough (turbulent and supersonic) to produce an explosive shock wave. The effect zone is determined by the quantity of energy released and the blast shock wave overpressure resulting from the explosion.

External Risk Reduction Facilities (ERRF) – measures to reduce or mitigate the risks which are separate and distinct from, and do not use, E/E/PE safety related systems or other technology safety related systems. For example, a drain system, fire wall, and bund are all external risk reduction facilities. (Source: IEC 61508.)

Fault Propagation Modeling – the analysis of the chain of events that leads to an accident. By analyzing what events initiate that chain, analyzing which events contribute to or allow the accident to propagate, and establishing how they are logically related, the event frequency can be determined. Fault propagation modeling techniques use the failure rates of individual components to determine the failure rate of the overall system.

Fault Tree Analysis – a top down approach to describing failures of complex systems. A fault tree analysis begins with the "top event," which is the result of a number of basic events that contribute to, or initiate, the system failure. The logic of a fault tree is indicated by the symbols representing the basic events and gates that logically relate those events.

Final Element (or actuator) – part of a safety instrumented system that implements the physical action necessary to achieve a safe state. (Source: IEC draft 61511.)

Financial Integrity Level (FIL) – when SIFs are used to decrease the likelihood or consequence of financial impact, a separate integrity level category can be assigned to the SIF as a result of these financial concerns. This Financial Integrity Level, or FIL, assignment is made using the same tools as SIL assignment (i.e., risk graph and hazard matrix), but the categories of consequence are described in terms of cost.

Fire (Flash) – the result of ignition of a cloud of flammable vapor, when the flame velocity is too slow (laminar and subsonic) to produce an explosive shock wave. When a gas phase mixture of fuel and air is ignited, a flame front travels from the point of ignition in all directions where the fuel/air concentration is within flammable limits. The velocity of the flame front will determine the type of damage that will be caused as a result of this event.

Fire (Jet) – results when high-pressure flammable material is ignited as it is being released from containment. The effect zone of a jet fire is proportional to the size of the flame generated. As a high-pressure material is released from a hole, the material will exit with a velocity that is mainly a function of system pressure and hole size. As distance away from the hole increases, the amount of oxygen in the mixture increases as air is entrained in the jet. As the upper flammability limit threshold is crossed, the fuel and air react, releasing the energy of combustion. As the combustion continues, entrained air, unburned fuel, and combustion products continue to move in the direction of the release due to the momentum generated by the release.

Fire (Pool) – results when spilled flammable liquids are ignited. The magnitude of the effect zone created by a pool fire will depend on the size of the flame that is generated, which in turn depends on the size of the spill surface and the properties of the spilled fluid. The flame's footprint is determined by the containment of the liquid spill, which is often controlled by any dikes or curbs present. If a spill is unconfined, the liquid will spread over an area determined by the fluid's viscosity and the characteristics of the surface on which the material is spilled, such as its porosity.

Fireball – result of a sudden and widespread release of a flammable gas or volatile liquid that is stored under pressure, coupled with immediate ignition. This is distinguished from a jet fire by the shorter duration of the event and the difference in the geometry and shape of the flame. When a pressure vessel containing a flammable gas or volatile liquid ruptures, the first result is the quick dispersion of the flammable material as the high-pressure material rapidly expands to atmospheric pressure. During this expansion, the release will entrain large quantities of air as a result of the process. If the material in the vessel is a volatile liquid, this process will also cause formation of an aerosol with the dispersion of liquid droplets away from the release as a result of the vapor expansion.

Functional Safety – part of overall safety relating to the equipment under control and its control system which depends on the correct functioning of the E/E/PE safety related systems, other technology safety related systems, and external risk reduction facilities. (Source: IEC 61508.)

Hazard – the potential for harm in a process. This is typically in the form of stored energy that can cause harm if control is lost. This stored energy could include high pressure, high temperature, energetic chemical reactivity (i.e., explosion or fire), and toxic reactivity. (Source: IEC 61508.)

Hazard Matrix – a qualitative and category-based method for assigning an SIL. The user must create a matrix that assigns broad categories to the consequence (one axis dimension) and likelihood (other axis dimension) components of the risk with an SIL assignment associated for each entry in the matrix. In some cases, quantitative tools, such as LOPA, are used to assist the analyst in determining which category to use, but often the assignment is done completely qualitatively, using engineering judgment.

HAZOP – a process hazards analysis procedure originally developed by ICI in the 1970s. The method is highly structured and divides the process into different operationally based nodes and investigates the behavior of the different parts of each node based on an array of possible deviation conditions or guidewords.

Incident – the result of an initiating event that is not stopped from propagating. The incident is the most basic description of an unwanted accident, and provides the least information. The term *incident* is simply used to convey the fact that the process has lost containment of the chemical or other potential energy source. Thus the potential for causing damage has been released, but its harmful result has not has not taken specific form. The CPQRA definition is: "The loss of containment of material or energy (e.g., leak of 10 lb/s of ammonia from a connecting pipeline to the ammonia tank, producing a toxic vapor cloud); not all events propagate into incidents."

Incident Outcome – the particular effect of the incident that is being analyzed. CPQRA defines incident outcome as: "The physical manifestation of the incident; for toxic materials, the incident outcome is a toxic release, while for flammable materials, the incident outcome could be a Boiling Liquid Expanding Vapor Cloud Explosion (BLEVE), flash fire, unconfined vapor cloud explosion, toxic release, etc. (e.g., for a 10 lb/s leak of ammonia, the incident outcome is a toxic release)."

Incident Outcome Case – a second level of classification for incident outcome. For instance, a release can occur under a variety of weather conditions, and each of those weather conditions will result in a different magnitude of consequence. As such, a number of different incident outcome cases are required to describe the same incident outcome. One incident outcome case would be used for each weather condition under study. CPQRA defines incident outcome case as: "The quantitative definition of a single result of an incident outcome through specification of sufficient parameters to allow distinction of this case from all others for the same incident outcomes. For example, a release of 10 lb/s of ammonia with a D atmospheric stability class and 1.4 mph wind speed gives a particular concentration profile, resulting, for example, in a 3000 ppm concentration at a distance of 2000 feet."

International Electrotechnical Commission (IEC) – a worldwide organization for standardization. The object of the IEC is to promote international cooperation on all questions concerning standardization in the electrical and electronic fields. To this end and in addition to other activities, the IEC publishes international standards. International, governmental, and non-governmental organizations liasing with the IEC also participate in this preparation. The IEC collaborates closely with the International Organization for Standardization (ISO), in accordance with conditions determined by agreement between the two organizations. (Source: IEC 61508.)

Layer of Protection Analysis (LOPA) – a method of fault propagation modeling closely related to the event tree method for determining the likelihood of a harmful incident outcome considering the different protection layers that could act to prevent it. The method starts with data developed in the process hazards analysis and accounts for each identified hazard by documenting the initiating cause and the protection layers that prevent or mitigate the hazard. The total amount of risk reduction can then be determined and the need for more risk reduction analyzed. If additional risk reduction is required and if it is to be provided in the form of a safety instrumented function (SIF), the LOPA methodology allows the determination of the appropriate safety integrity level (SIL) for the SIF. (Source: IEC draft 61511.)

Likelihood – the frequency of a harmful event often expressed in events per year or events per million hours. One of the two components used to define a risk.

Markov Analysis – a fault propagation method used to analyze failure rate or probability for very complex systems such as fault tolerant PLC. A diagram is constructed to represent the system under consideration, including the logical relationships between its components. In Markov analysis there is a group of circles, each of which represents a system state. The different states are connected with transitions, which are shown as arrows and indicate paths to move from one state to another. The transitions are quantified using either failure rates when the transition is from an OK state to a failed state or repair rates when the transition is from a failed state back to an OK state. As with other models, the diagram is converted into a set of mathematical equations that are then solved to quantify the likelihood of the event or state in question.

Nuisance Trip (Safe Failure) – failure that does not have the potential to put the safety instrumented system in a dangerous or fail-to-function state. The situation when a safety related system or component fails to perform properly in such a way that it calls for the system to be shut down or the safety instrumented function to activate when there is no hazard present. (Source: IEC draft 61511.)

Occupancy – a measure of the probability that the effect zone of an accident will contain one or more personnel receptors of the effect. This probability should be determined using plant-specific staffing philosophy and practice.

OSHA PSM (Occupational Safety and Health Administration Process Safety Management) – the US regulation (29 CFR 1910.119) published in 1992 to help prevent or minimize the consequences of catastrophic releases of toxic, reactive, flammable, or explosive chemicals. Its stated purpose is to accomplish this goal by requiring a comprehensive management program: a holistic approach that integrates technologies, procedures, and management practices. This includes the requirement to conduct a formal process hazards analysis. The regulation applies to processes that involve chemicals at or above threshold quantities (specified in Appendix A of the standard) and processes that involve flammable liquids or gases on-site in one location, in quantities of 10,000 pounds or more (subject to few exceptions). Hydrocarbon fuels, which may be excluded if used solely as a fuel, are included if the fuel is part of a process covered by this standard. (Source: OSHA.)

Physical Relief Device – mechanical equipment that performs an action to relieve pressure when the normal operating range of temperature or pressure has been exceeded. Physical relief devices include pressure relief valves, thermal relief valves, rupture disks, rupture pins, and high-temperature fusible plugs.

Probability of Failure on Demand (PFD) – the probability that a system or component will fail to execute the task for which it was designed within an appropriate time period when it is called upon to do so.

Probable Loss of Life (PLL) – a numerical expression for the magnitude of a consequence in terms of the most likely number of lives that will be lost in a given event or over a given time interval. The value need not be a whole number.

Process Hazards Analysis – the step in the safety life cycle (required by OSHA PSM where applicable) to identify potential process hazards including their consequence and likelihood. The report generated by this effort is useful in identifying existing and recommended safety instrumented functions. The IEC 61508 requires that the PHA determines the hazards and hazardous events of the equipment under control and its control system (in all modes of operation), for all reasonably foreseeable circumstances including fault conditions and misuse. The OSHA PSM requires the PHA to address the hazards of the process, previous hazardous incidents, engineering and administrative controls, the consequences of the failure of engineering and administrative controls, human factors, and an evaluation of the effects of failure of controls on employees. (Source: IEC 61508 and OSHA.)

Proof Test – periodic test performed to detect failures in a safety related system so that, if necessary, the system can be restored to an "as new" condition or as close as practical to this condition. (Source: IEC 61508.)

Protection Layer – any independent mechanism that reduces risk by control, prevention, or mitigation. This could include: a process engineering mechanism such as the size of vessels containing hazardous chemicals, etc., a mechanical engineering mechanism such as a relief valve, a safety instrumented system, or an administrative procedure such as an emergency plan against an imminent hazard. These responses may be automated or initiated by human actions. (Source: IEC draft 61511.)

Receptor – the object or persons on the receiving end of the harm in an unwanted event. Common receptors include personnel, plant equipment, plant production, the environment, and the general public.

Risk – combination of the probability of occurrence of harm and the severity of that harm. (Source: IEC 61508.)

Risk (Geographic) – a measure of the probability that an event will occur in a specific geographic location. Geographic risks are typically shown by drawing lines of constant risk (isopleths) on a process plot plan.

Risk (Individual) – the frequency at which an individual may be expected to sustain a given level of harm from the realization of specified hazards. Individual risk is typically measured as a probability of fatality per year. For process plants this level of risk is typically determined for the maximally exposed individual. In practice it is important to clarify whether this refers to exposure time or total time since the numbers are distinctly different.

Risk (Societal) – the relationship between the frequency and the number of people suffering from a specified level of harm. This is typically shown as a plot of number of fatalities versus the cumulative frequency of events on a log-log chart, commonly referred to as an F-N curve, or Farmer curve.

Risk (Tolerable) – the non-zero level of risk that is acceptable to an organization. This level helps to identify how much risk reduction is required to be provided by any safety instrumented system installed as part of a process. This acceptable level must be defined in terms of both consequence and likelihood.

Risk (Unmitigated) – the level of risk that is present in a process before any safety instrumented systems are considered. This level helps to identify how much risk reduction is required to be provided by any safety instrumented system installed as part of a process. This unmitigated risk level must be defined in terms of both consequence and likelihood.

Risk Aversion – the concept where, given two sets of events that cause the same number of fatalities, the set of events that causes fewer fatalities per event is more acceptable to society. This is often included in risk analysis by raising the probable loss of life term to a power of a risk aversion factor "α" in assessing the magnitude of the consequence. A risk aversion factor of 1 implies risk neutrality, where the amount of harm per incident in relation to the total harm is not relevant. A risk aversion factor of 2, on the other hand, would imply one order of magnitude of risk aversion, meaning that a single event with ten fatalities would be an order of magnitude less tolerable than ten individual fatality events.

Risk Graph – a qualitative and category-based method of SIL assignment. Risk graphs were initially developed for German industry standards and are commonly used in the European Community. Risk graph analysis uses four parameters to make an SIL selection: consequence, occupancy, probability of avoiding the hazard, and demand rate. Each of these parameters is assigned a category and an SIL is associated with each combination of categories. In some cases, quantitative tools, such as LOPA, are used to assist the analyst in determining which category to use, but typically the assignment is done qualitatively. Using the selected categories, the analyst follows the resulting path that leads to the associated SIL assignment.

Risk Integral – a summation of risk as expressed by the product of consequence and frequency. The integral is summed over all of the potential unwanted events that can occur. If calculating the risk integral for loss of life, the consequence of concern and thus the units of the integral are fatalities.

Safe State – the state of the equipment under control when there is freedom from unacceptable risk. (Source: IEC 61508.)

Safety Instrumented Function (SIF) – a function with a specified safety integrity level, which is intended to achieve or maintain a safe state for the process with respect to a specific hazardous event. (Source: IEC draft 61511.)

Safety Instrumented System (SIS) – a set of sensors, logic solvers, and actuators designed to carry out one or more safety instrumented functions. (Source: IEC draft 61511.)

Safety Integrity Level (SIL) – discrete level (one out of a possible four) for specifying the probability of a safety instrumented system satisfactorily performing the required safety instrumented functions under all of the stated conditions within a stated period of time. Safety integrity level 4 has the highest level of safety integrity and safety integrity level 1 has the lowest. (Source: IEC 61508 and draft 61511.)

Safety Integrity Level (SIL) Assignment – the final step of SIL selection in which the SIL is assigned to a specific SIF based on the difference between the unmitigated risk for the process in question and the risk that is tolerable to the owner of the process.

Safety Integrity Level (SIL) Selection – the overall process of identifying potential safety instrumented functions, assessing the likelihood and consequence of the unmitigated risk of the harmful events they prevent, and assigning a safety integrity level based on the risk reduction required to make the risk tolerable when implementing a safety instrumented system.

Safety Life Cycle (SLC) – necessary activities involved in the implementation of safety instrumented functions, occurring during a period of time that starts at the concept phase of a project and finishes when all of the safety instrumented functions are no longer available for use. (Source: IEC draft 61511.)

Safety Related System – designated system that both: implements the required safety functions necessary to achieve or maintain a safe state for the equipment under control, and is intended to achieve—on its own or with other E/E/PE safety related systems, other technology safety related systems, or external risk reduction facilities—the necessary safety integrity for the required safety functions. (Source: IEC 61508.)

Safety Requirements Specification (SRS) – specification containing all of the requirements of the safety functions that have to be performed by the safety related systems. (Source: IEC draft 61511.)

Sensor – device or combination of devices that measures the process condition (e.g., transmitters, transducers, process switches, position switches, etc.). (Source: IEC draft 61511.)

System – set of elements that interact according to a design. An element of a system can be another system and may include hardware, software, and human interaction. (Source: IEC draft 61511.)

Vulnerability – the probability that a receptor, whether human or equipment, will sustain a defined level of harm given that it has been exposed to the effect of an incident. For instance, if a person is in a normally occupied building that sustains a collapse, CCPS estimates that he or she has a fatal injury vulnerability of 0.6.

APPENDIX D

Problem Solutions

The solutions to the sample problems at the end of each chapter are shown below.

Chapter 1

1.1 For SIL 2, the PFD range is 10^{-2} to 10^{-3}. For SIL 1, the RRF range is 10 to 100. Risk reduction factor is the inverse of probability of failure on demand (i.e., RRF = 1/PFD).

1.2 The standards that describe the safety life cycle include: ANSI/ISA-84.01-1996, IEC 61508, and IEC draft 61511.

1.3 According to IEC draft 61511-1, clause 3.2.66 and 69, a safety instrumented function is a "safety function with a specified safety integrity level which is necessary to achieve functional safety," and as a safety function is "to be implemented by a safety instrumented system, other technology safety related system or external risk reduction facilities which is intended to achieve or maintain a safe state." Thus an SIF is a *single* action. In practice, an SIF also can refer to the set of sensors, logic solver, and actuators that performs that *single* specific function and protects against one specific hazard. The safety instrumented system is then a group of safety instrumented functions that are logically related by project, process unit, or other logical grouping.

1.4 Likelihood and consequence.

1.5 The anchoring trap – Over-relying on first thoughts.

The status quo trap – Focusing on the current design.

The sunk-cost trap – Protecting earlier choices.

The recallability trap – Focusing on dramatic events.

The base-rate trap – Neglecting relevant information.

The prudence trap – Slanting probabilities and estimates.

1.6 Layer of protection analysis is a method for estimating the likelihood, or frequency, of an unwanted event.

Chapter 2

2.1 The main IEC 61508 SLC analysis phase steps are: 1-Concept, 2-Overall Scope Definition, 3-Hazard and Risk Analysis, 4-Overall Safety Requirements, and 5-Safety Requirements Allocation. Thus the process of SIL selection is included in this phase.

2.2 Mistakes in the analysis phase, which includes system specification, cause the most harmful accidents at 44% of the total.

2.3 A good SRS should first and foremost include the overall safety requirements in terms of listed and defined safety instrumented functions with their SIL requirements.

2.4 The testing philosophy is considered at this early phase because the test interval can change the SIL of a given set of equipment. More frequent proof testing can increase the SIL in certain cases.

2.5 The function of the physical SIS is tested during the validation step.

2.6 Maintenance and function testing dominate the operation phase of the SLC.

Chapter 3

3.1 ALARP stands for As Low As Reasonably Practicable and means that above a certain acceptable threshold, risk should be reduced as much as it is practical to do so. The three levels of risk are "Unacceptable" where the risk should not be undertaken at all, the "ALARP" region where practical risk reduction is required, and the "Broadly Acceptable" region where the risk is already acceptable and no further reduction needs to be considered.

3.2 One method is calculating the future earnings of the individual as a measure of the value of a saved life. Another method is determining society's "willingness to pay" for risk reduction by analyzing revealed values associated with a saved life. Another method is based on the historical data from awards and costs associated with wrongful death lawsuits.

3.3 Measures of risk include: individual, societal, and geographic risk as well as risk integrals and expected value.

3.4 Taking maximum individual risk criteria of 4.5×10^{-5} per year and dividing by the total number of fatalities gives the tolerable event frequency of 2.2×10^{-6} per year. Note that this assumes there is no risk aversion factor associated with events that produce a larger number of fatalities.

3.5 Taking the event frequency of 3.9×10^{-4} and multiplying by the probable loss of life of 0.1 (after rearranging equation 3.2 for F_{IND}, or the individual risk of fatality) gives an observable lower bound of 3.9×10^{-5} for the plant's tolerable individual risk of fatality.

3.6 Risk reduction projects are cost effective for benefit-to-cost ratios of more than one.

3.7 Taking the 2,000 hours per year in the environment of V-101/102 in figure 3.4 with no other risks gives an average annual risk of between 10^{-3} and 10^{-4} times 2,000/(24x365), which translates into between 2.3×10^{-4} and 2.3×10^{-5}.

3.8 Annual cost of accident without protection = $35,000,000/1,900 = $18,421.
Annual cost of accident with protection = 0.95 × $5,000,000/1,900 + 0.05 × $35,000,000/1,900 = $2,500 + $921 = $3,421.
Gross benefit of fire water system = $18,421-$3,421 = $15,000.
Benefit-to-cost ratio = $15,000/$5,930 = 2.53
This is greater than one, indicating that the fire system is a wise investment.

3.9 False. The United States specifically does not set an individual risk of fatality threshold for either workers or the general public.

Chapter 4

4.1 Sources of information to identify existing or recommended SIFs include: process hazards analysis reports, piping and instrumentation diagrams, process flow diagrams, and other engineering drawings and documentation.

4.2 Methods to perform process hazards analysis include: checklist, "what if" study, failure modes and effects analysis, fault tree, and HAZOP.

4.3 SIF: Transfer pump shut-off on high feed tank level.
Initiating Event: Delivery truck contains too much material for the tank.
Consequence: Exposure to toxic gas release.
Safeguard: Alarm and operator action to shut off pump.

4.4 According to IEC draft 61511-1, clause 3.2.66 and 69, a safety instrumented function is a "safety function with a specified safety integrity level which is necessary to achieve functional safety," and as a safety function is "to be implemented by a safety instrumented system, other technology safety related system or external risk reduction facilities which is intended to achieve or maintain a safe state." Thus an SIF is a *single* action. In practice, an SIF also can refer to the set of sensors, logic solver, and actuators that performs that *single* specific function and protects against one specific hazard.

Chapter 5

5.1 Probability multiplication is used to calculate the probability of a logical AND of two independent events.

5.2 The two methods of probability assignment are: 1) analysis of the physical properties and characteristics surrounding the event, and 2) empirical analysis of historical or experimental data. The probability of drawing the ace of spades from a deck of cards is most easily determined by analysis of the physical properties of the event.

5.3 The probability of failure of a relief valve with a 99% success probability is 100% - 99% = 1%.

5.4 Probability multiplication is used based on the logical AND to give: $P_{\text{Overfill}} = P_{\text{Big truck load}} \times P_{\text{Operator error}} = 0.25 \times 0.1 = 0.025$.

5.5 As shown in the table below, there are three possible combinations that give a roll of 4 out of the 36 total combinations possible with two six-sided dice. Thus the probability of rolling a 4 is 3/36 = 0.0833.

Die 1	Die 2
1	3
3	1
2	2

5.6 The fault tree in question starts with an AND condition, between the two power supply failure basic events drawn as circles in the lower left of the diagram, that is required for the case where the valve closes because of a power failure. Then the fault tree links the result of the power failure AND gate to the valve stem breakage basic event with an OR gate to generate the top event where the valve fails closed. Assuming all of these events are independent but not mutually exclusive leads to the calculation based on equations 5.4 and 5.7.

$P_{\text{Valve Fails}} = 1-(1-(0.2 \times 0.05))\times(1-0.01) = 1-0.99\text{x}0.99 = 0.0199$

Using the simplified equation 5.5 yields:

$P_{\text{Valve Fails}} = (0.2 \times 0.05) + 0.01 = 0.02$, which is not much different from the exact answer.

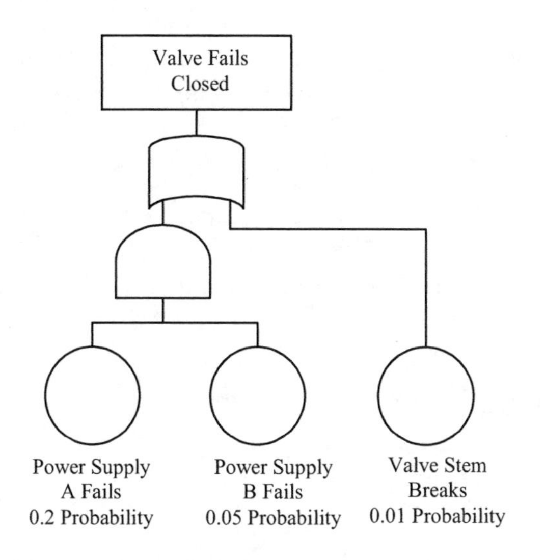

5.7 The total number of valve years of operation is 200 × 15 = 3,000 while the total number of failures is 75. Thus the failure rate of the valves is F = 75/3,000 = 0.025 failures per valve per year. Then using equation 5.11 with the 1 year proof test interval, one gets $PFD_{Max} = 1 - e^{(-0.025 \times 1)} = 0.0247$. Or with the simplified equation 5.16, one gets $PFD_{Max} = 0.025 \times 1 = 0.025$ which is quite close to the exact answer.

5.8 The MTTF calculation of a component with a $PFD_{Avg} = 0.05$ and a 3 year test interval starts with the simplified equation 5.13 rearranged to solve for the failure rate $\lambda = PFD_{Avg} \times (2/t) = 0.05 \times 2/3 = 0.0333$ per year. Then taking the inverse with equation 5.10 gives an MTTF of 30 years.

5.9 Plant-specific database records provide a more accurate failure rate prediction because they inherently take into account the particular environment under which the equipment will be operating and the way in which it will be maintained. Both of these factors can significantly affect failure rates, thus making any generic data less relevant and thus less accurate.

Chapter 6

6.1 Semi-quantitative consequence analysis tools include the Dow Fire and Explosion Index and the Dow Chemical Exposure Index. These methods use general process parameters to give the process a *score* that reflects a relative level of risk. The semi-quantitative aspects stem from the numerical *score* output being difficult to translate into an actual zone that would be impacted by an accident. Thus, although a number is used in the analysis, it includes some qualitative aspects in translating the result into the context of a full SIL selection process.

6.2 Physical explosions derive their primary explosive energy from the compression work that is released when containment is lost. Vapor cloud explosions derive their primary explosive energy from the heat of reaction given off during the explosion. In both cases, the energy is converted to a shock wave that can subsequently damage objects in its path.

6.3 The consequence is determined directly by taking the statistical average boiler explosion result of 12/1,012 = 0.0119 fatalities per explosion event.

6.4 Since there is no information to the contrary, the effect zone is assumed to be circular with the 25-meter distance as the radius. Thus the area $= \pi \times r^2 = 3.14 \times (25)^2 = 1963\ m^2$.

6.5 A flammable gas release can result in a pool fire if the thermodynamics of the material and the ambient conditions combine to cause the vapor to condense into a liquid after it is released. This "rain out" can then create a pool of flammable liquid some distance from the release, which could then ignite and cause a pool fire.

6.6 Factors that determine the size of the effect zone of a toxic release include: release quantity, duration of release, source geometry, elevation/orientation of the release, initial density of the release, prevailing atmospheric conditions, surrounding terrain, and limiting concentration (endpoint).

6.7 Consequence-modeling tools that perform computer calculations of toxic gas effect zones include: FETCH™, ARCHIE™, ALOHA™, DEGADIS™, PHAST™, and HGSYSTEM™.

6.8 The average capital density is calculated by taking the total capital cost divided by the total plant area. Thus €25,000,000/(75x84) = €3,968 per m^2.

6.9 The control room occupancy is calculated by summing up the contributions from each work group:
Engineers = 20 × 0.95 = 19.0 persons
Supervisor = 1 × 0.80 = 0.8 persons
Maintenance = 40 × 0.10 = 4.0 persons
Total Occupancy = 19.0 + 0.8 + 4.0 = 23.8 persons

6.10 The PLL consequence is equal to the personnel density times the fatality effect zone times the vulnerability, so PLL = 0.002 × 5600 × 100% = 11.2. The EV consequence is equal to the capital density times the equipment damage effect zone times the vulnerability, so EV = ¥150,000 × 2400 × 100% = ¥360,000,000.

Chapter 7

7.1 The primary fault propagation models include: event tree analysis, reliability block diagrams, fault tree analysis, and Markov analysis. All of these models are based on constructing a diagrammatic representation of the events and interconnecting logic that lead to system failure. These diagrams then allow mathematical equations to be applied to the systems to determine the ultimate failure rate or likelihood in question. The primary differences are based on their complexity and their optimal applications. Event tree analysis is usually the most straightforward and easy-to-use modeling technique for likelihood analysis, although its application is limited because it cannot analyze complex situations very well. Reliability block diagrams have the strength of being easy to use, but are limited in practice to situations where the governing equations have already been derived. Fault tree analysis can address more complex logic than either event tree analysis or reliability block diagrams, but it is still limited by the need to use pre-derived equations to calculate the failure rates of basic events. Markov analysis is the most complex of the methods mentioned here, but it is also the most accurate and flexible technique generally used for fault-tolerant PLCs and similar equipment.

7.2 Fault propagation modeling is preferred over statistical methods for process plant failures when the plant is new, different, or complex, because there is typically not enough relevant statistical history applicable at the whole plant level. In those cases, the fault propagation methods take advantage of the similarity at the more detailed level. In this way, more reliable failure rate data on the individual components of the plant can be used with fault propagation modeling to calculate more accurate whole-plant estimates.

Chapter 8

8.1 The event tree is shown in the figure below. Assume an initiating event frequency of 52 times per year to give the starting frequency at the left, since no information is given in the problem. The outcome frequencies are calculated by using probability multiplication along each chain of interest to give the values shown in the spreadsheet.

INIT EVENT	BRANCH 1	BRANCH 2	BRANCH 3	OUTCOME
Material	Not enough	Operator does	Operator does	
Delivery	room in tank	not notice	not detect high	
		initial level	level during load	
			15%	0.039/yr
			TRUE	Material spill
		5%		
		TRUE	85%	0.221/yr
			FALSE	Stopped during fill
	10%			
52 per year	TRUE	95%		4.94/yr
		FALSE		Stopped before fill
	90%			46.8/yr
	FALSE			No fill problem

8.2 The event tree is shown in the following figure. The event probabilities are noted at each branch and probability multiplication is used to calculate the contribution of each potential outcome to the average consequence based on its frequency or probability weighting. The contributions are then summed to give the average result.

INIT EVENT	BRANCH 1	BRANCH 2	OUTCOME	EV	CONTRIB.
Gas	Ignition	Explosion			
Release					
		15%	0.0375	$45,000,000	$1,687,500
	25%	TRUE	VCE		
	Delayed				
1		85%	0.2125	$850,000	$180,625
		FALSE	Flash Fire		
	5%		0.05	$5,000,000	$250,000
	Immediate		Jet Fire		
	70%		0.7	0	$0
	None				
			Average Consequence in terms of EV:		$2,118,125

Chapter 9

9.1 Since the outcome is linked to the initiating and intermediate events by logical ANDs, probability multiplication is used to calculate LOPA frequencies.

9.2 The primary difference between LOPA and event tree analysis is that LOPA only considers the single chain of events that leads to a harmful outcome while event trees consider all possibilities and multiple outcomes. In LOPA, all branches are complementary events representing the failure of a protection layer. In event tree analysis, branches are not limited to complementary events and can represent a much wider range of event types than just protection layer failures.

9.3 A formal protection layer is a specifically engineered device such as a pressure relief valve or a high-temperature shut-off switch, while a mitigating event is something that reduces risk but is not specifically designed to do so, such as probability of ignition and probability of explosion.

9.4 The criteria for a formal protection layer are:
Specificity – An independent protection layer must be specifically designed to be capable of preventing the consequences under consideration.
Independence – The operation of the protection layer must be completely independent from all other protection layers; no common equipment can be shared with other protection layers.
Dependability – The device must be able to dependably prevent the consequence from occurring. Both systematic and random faults need to be considered in its design.
Auditability – The device should be proof tested and maintained. These audits of operation are necessary to ensure that the specified level of risk reduction is being achieved.

9.5 Common protection layers and mitigating events include: basic process control system (BPCS), operator intervention, use factor, mechanical integrity of vessel, physical relief device, external risk reduction facilities, ignition probability, explosion probability, and occupancy.

9.6 The mechanical integrity layer of protection PFD is approximated by the chance of the vessel failing during one year of normal service. This is calculated using equation 5.11 so
$PFD = 1 - e^{(-1.1 \times 10\text{-}7 \times 8760)} = 1 - 0.99904 = 0.00096$

9.7 The LOPA diagram for this problem is shown in the following figure. The initiating event is the cooling water pump failure and its frequency is listed at the left in the diagram. The rupture disk PFD is entered directly, while that for the relief valve must be calculated. This requires an assumption of a proof test interval. In this case, one year is chosen. Using equation 5.11, PFD = $1 - e^{(-4.5 \times 10\text{-}6 \times 8760)} = 1 - 0.961 = 0.039$, so this value is added to the diagram for the relief valve. Using the information in table 9.1, one determines that the operator falls into the "Normal" operator response category since the response is trained but not drilled repeatedly and a PFD of 0.1 is entered. This then gives the overall frequency from probability multiplication of 1.8×10^{-9} per hour. Converting this figure into frequency per year, which is more useful for comparison with risk tolerance targets, yields 1.6×10^{-5}.

INIT EVENT	PL #1	PL #2	PL #3	OUTCOME
Loss of	Rupture	Relief	Plant	Outcome
Cooling Water	Disk	Valve	Operator	
				1.76E-09
			0.1	Vessel
		0.039		Rupture
	0.10%			
4.5x10-4				
failures per hour				No Event

9.8 Looking at the information in table 9.1, the operator meets all of the "Normal" operator response criteria: ample indications a condition requiring a shutdown exists, operator has been trained in proper response, operator is always monitoring the process (relieved for breaks), EXCEPT for the last one—operator has ample time (> 20 minutes) to perform the shutdown. Since no information is provided in this final area, the "Response unlikely" category must be chosen with a PFD of 1.0. Note that this could be improved to the "Normal" operator response of PFD = 0.1 if the information were available and it clearly satisfied the remaining condition.

9.9 The use factor for a LOPA in this case is calculated from the total cycle time, since the batch process is run without any down time. The total cycle lasts 10 hours, while the potential for release only occurs during the 1.5-hour reaction phase. Taking the ratio, one gets a use factor of 0.15.

Chapter 10

10.1 Since SIL 1 systems can have a risk reduction factor of anywhere from 10 to 100, it is not a sufficient single specification. SIL 2 must be selected to guarantee that the risk reduction factor is greater than 50. While not formally recognized by industry group standards, some organizations allow non-integer SIL assignments or SILs with RRF assignments; in these cases the assignment would be SIL 1.7 or SIL 1 with RRF 50, respectively.

10.2 According to table 10.2, a failure rate of 5.0×10^{-6} per hour falls into the 10^{-5} to 10^{-6} failure per hour range of SIL 1.

10.3a A multiple fatality consequence of 7 deaths translates into an "Extensive" category while a remote likelihood of 10^{-6} per year translates into a "Low" category. Reading off the corresponding SIL from the hazard matrix in figure 10.1 gives 3*, which means one SIL 3 system may be insufficient for the task. In this case, it may pay to investigate other means of risk reduction in addition to an SIS.

10.3b A consequence of minor injury translates into a "Minor" category while a remote likelihood of 10^{-6} per year translates into a "Low" category. Reading off the corresponding SIL from the hazard matrix in figure 10.1 gives NR, which means no risk reduction is required for the situation under consideration.

10.3c A consequence of a single fatality translates into a "Serious" category while a likelihood of 1/25 per year translates into a "High" category. Reading off the corresponding SIL from the hazard matrix in figure 10.1 gives 3*, which means one SIL 3 system may be insufficient for the task. In this case, it may pay to investigate other means of risk reduction in addition to an SIS.

10.4a The multiple fatalities, remote likelihood, no probability of avoidance, and not normally occupied area translate into a C_D, W_1, P_B, and F_A set of categories. This gives row X_5 of column W_1, which indicates a SIL of 2 in figure 10.2.

10.4b The minor injury, remote likelihood (~10^{-6} per year), good possibility of avoidance, and normally occupied area translate into a C_A, W_1, P_A, and F_B set of categories. This gives row X_1 of column W_1, which indicates there are no safety requirements in figure 10.2.

10.4c The single fatality and moderate likelihood (~1/25 per year), are the only data given, so no other credit can be taken. This translates into a C_C, W_2, P_B, and F_B set of categories. This gives row X_5 of column W_2, which indicates a SIL of 3 in figure 10.2.

Figure 10.1 Typical Hazard Matrix

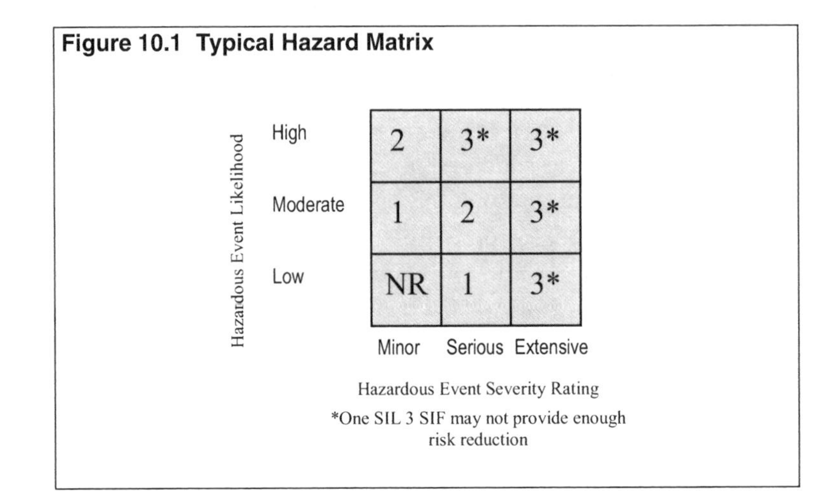

Figure 10.2 Typical Risk Graph

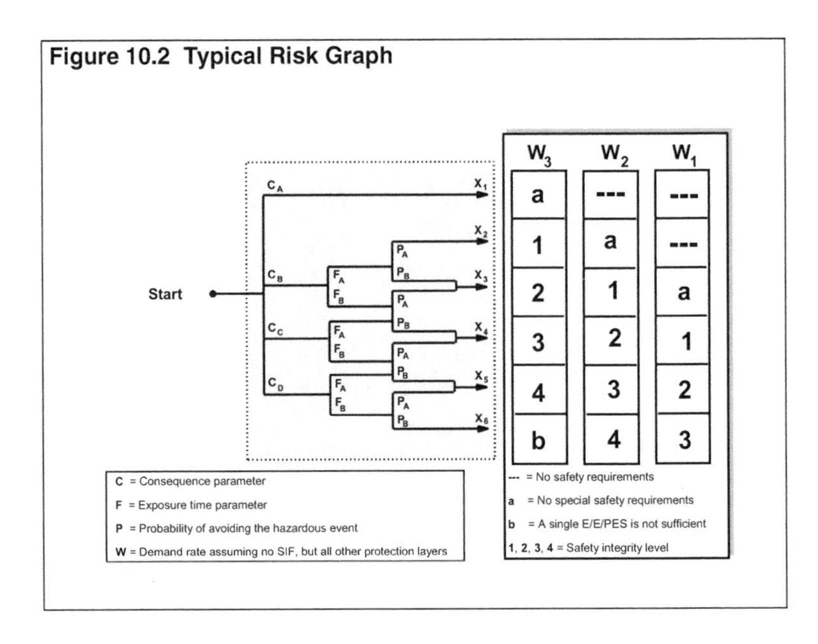

10.4d The precision of the probable loss of life of 0.98 estimate indicates that it was calculated using either consequence modeling or statistical means. These advanced methods of consequence estimation incorporate occupancy and probability of avoidance. Therefore, no additional credit should be taken in the risk graph. Combined with a high likelihood of 0.45 times per year, this translates into a C_C, W_3, P_B, and F_B set of categories. This gives row X_5 of column W_3, which indicates a SIL of 4 in figure 10.2. With such a high SIL, it may be advised to look into other additional methods of risk reduction.

10.4e The $7 million loss and low likelihood (~1/700 per year), are the only data given so no other credit can be taken. Using table 10.3 in comparison with table 10.5 indicates that a $7 million loss is probably most closely represented by category C_C. Overall, this then translates into a C_C, W_1, P_B, and F_B set of categories. This gives row X_5 of column W_1, which indicates a SIL of 2 in figure 10.2.

10.5a An extensive consequence in table 10.9 has a maximum tolerable frequency of 10^{-7} per year. So with an unmitigated likelihood of 3.4×10^{-3} per year, the required risk reduction is the ratio of the two numbers or 34,000. Since this is above 10,000, a SIL-only specification is not valid since more than a SIL 4 is required. The combined specification of SIL 4 with an RRF of >34,000 is acceptable but other methods should also be considered.

10.5b A minor consequence in table 10.9 has a maximum tolerable frequency of 10^{-4} per year. So with an unmitigated likelihood of 9.7×10^{-7} per year the required risk reduction is the ratio of the two numbers or 9.7×10^{-3}, which means no risk reduction, or SIL is required. It should be noted that calculations should not have been required to perform this assignment. Inspection of the tolerable event frequency and unmitigated event frequency should have revealed that no SIS is required because the unmitigated event frequency is less than the tolerable event frequency. If this is not noted prior to performing calculations, the results will show a required PFD greater than one and a required RRF of less than one, which are also indications that the risk is tolerable without any further mitigation.

10.5c A serious consequence in table 10.9 has a maximum tolerable frequency of 10^{-5} per year. So with an unmitigated likelihood of 8.0×10^{-2} per year, the required risk reduction is the ratio of the two numbers or 8,000. Since this is between 1,000 and 10,000, a SIL-only specification of SIL 4 is required. The combined specification of SIL 3 with an RRF of >8,000 is acceptable but other methods should also be considered.

10.6 With a maximum individual risk of fatality criterion of 2.5×10^{-5} per year, a risk aversion factor of 1, and a consequence with a PLL of 2.4, equation 10.5 gives a tolerable event frequency of 2.5×10^{-5} per year / 2.4 = 1.04×10^{-5} per year. Comparing this to the unmitigated frequency of 1/180 per year = 5.56×10^{-3} per year gives a ratio of 534 which corresponds to the risk reduction factor. A SIL only specification should then be SIL 3 to insure the required RRF is always met by the SIL.

10.7 Given that the maximum individual risk of fatality criterion is 1.0×10^{-5} per year and that there is risk neutral approach to societal risk (i.e., risk aversion factor of one), one then uses those criteria with the PLL after taking all of the parameters into account to calibrate the graph. Either a prototypical or a minimum value calibration can be performed. This solution will use the minimum value approach.

It is often easiest to start with a final effective maximum consequence parameter PLL of 1 and find the SIL for that row of the graph, since the event frequency will exactly match the individual risk tolerance frequency. In this case, that row would be C_C, F_B, and P_B, which is designated as X_5. Note that the F_B, P_B values do not change the effective maximum consequence parameter since their values are 1.0. The worst-case frequencies W_1, W_2, and W_3 are then correspondingly 0.1, 1, and 10 per year. The risk reduction factors to bring these frequencies below the maximum individual risk of fatality criterion of 1.0×10^{-5} per year are then 10,000, 100,000, and 1,000,000. This translates into SIL requirements of 4, b, and b, where b indicates that a single E/E/PES is not sufficient to reduce the risk to a tolerable level. These values are then entered into the corresponding boxes in the risk graph.

Next, using the same method for the effective maximum consequence parameter PLL of 10, the row would be C_D, F_B, P_B designated as X_6. The worst-case frequencies W_1, W_2, and W_3 are then correspondingly 0.1, 1, and 10 per year. Now since the PLL is 10, rather than 1, the event frequency must be calculated from equation 10.5, which gives a maximum value of 1.0×10^{-6} per year. The risk reduction factors to bring the W_1, W_2, and W_3 frequencies below the maximum event frequency of 1.0×10^{-6} per year are then 100,000, 1,000,000, and 10,000,000. This translates into SIL requirements of b, b, and b, where b again indicates that a single E/E/PES is not sufficient to reduce the risk to a tolerable level.

Using the same method for the effective maximum consequence parameter PLL of 0.1, the row would be C_B, F_B, P_B designated as X_4. The worst-case frequencies W_1, W_2, and W_3 are then correspondingly 0.1, 1, and 10 per year. With a PLL of 0.1, the event frequency from equation 10.5 is 1.0×10^{-4} per year. The risk reduction factors to bring the W_1, W_2, and W_3 frequencies below the maximum event frequency of 1.0×10^{-4} per year are then 1,000, 10,000, and 100,000. This translates into SIL requirements of 3, 4, and b, where b again indicates that a single E/E/PES is not sufficient to reduce the risk to a tolerable level.

Using the same method for the effective maximum consequence parameter PLL of 0.01, the row would be C_A, designated as X_1. The worst-case frequencies W_1, W_2, and W_3 are then correspondingly 0.1, 1, and 10 per year. With a PLL of 0.01, the event frequency from equation 10.5 is 1.0×10^{-3} per year. The risk reduction factors to bring the W_1, W_2, and W_3 frequencies below the maximum event frequency of 1.0×10^{-4} per year are then 100, 1,000, and 10,000. This translates into SIL requirements of 2, 3, and 4.

With rows X_2 and X_3 remaining, they are calculated in a similar way but the F and P parameters are no longer 1.0. For row X_2, parameters C_B, F_A, P_A apply which gives an effective maximum consequence parameter PLL of $0.1 \times 0.1 \times 0.1 = 0.001$. The worst-case frequencies W_1, W_2, and W_3 are then correspondingly 0.1, 1, and 10 per year. With an effective PLL of 0.001, the event frequency from equation 10.5 is 1.0×10^{-2} per year. The risk reduction factors to bring the W_1, W_2, and W_3 frequencies below the maximum event frequency of 1.0×10^{-2} per year are then 10, 100, and 1,000. This translates into SIL requirements of 1, 2, and 3. For row X_3, parameters C_B, F_A, P_B apply which gives an effective maximum consequence parameter PLL of $0.1 \times 0.1 \times 1 = 0.01$. The worst-case frequencies W_1, W_2, and W_3 are then correspondingly 0.1, 1, and 10 per year. With an effective PLL of 0.01, the event frequency from equation 10.5 is 1.0×10^{-3} per year. The risk reduction factors to bring the W_1, W_2, and W_3 frequencies below the maximum event frequency of 1.0×10^{-3} per year are then 100, 1,000, and 10,000. This translates into SIL requirements of 2, 3, and 4, thus completing the risk graph calibration.

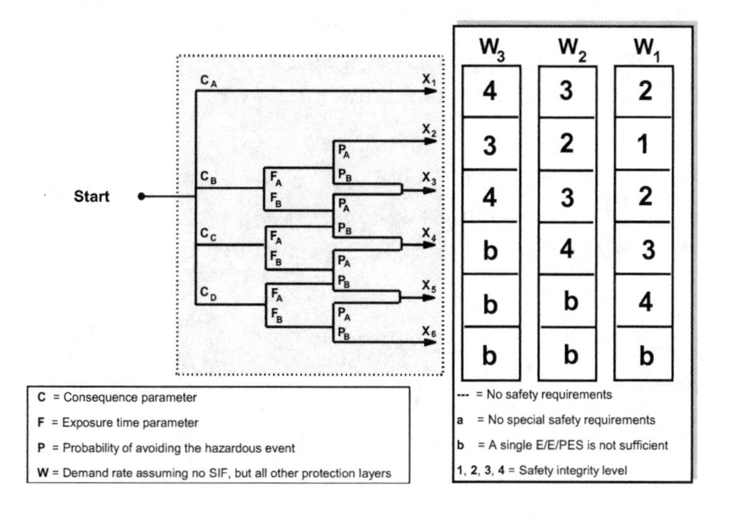

10.8 The starting point to reverse engineer the given risk matrix into a maximum individual risk of fatality and risk aversion factor is characterizing the PLL=1 case and applying the result to equation 10.3. This negates any effect of the risk aversion factor since one raised to any power is still one. Using the minimum value approach, the consequence is extensive and at the highest frequency of 1 event per year, the recommended SIL is 4, which translates into a minimum risk reduction factor of 10,000. Taking the ratio of the risk reduction factor to the frequency parameter according to equation 10.3, one gets a maximum mitigated event frequency of 10^{-4} per year for a single fatality.

Thus the intermediate state of equation 10.3 is

$$f_{Tolerable} = \frac{f_{Tolerable,\ Individual}}{PLL^{\alpha}} = \frac{1.0 \times 10^{-4}}{PLL^{\alpha}}$$

To find the value for the risk aversion factor α, one needs to examine a case with a PLL of greater than 1, since it is not appropriate to use the risk aversion factor for events with less than one fatality. In this problem, there is no such case, so risk aversion is not considered in the calibration. When risk aversion is not considered, risk neutrality or $\alpha=1$ is used. Substituting $\alpha=1$ and PLL=1 into the equation above yields a maximum individual risk of fatality target of 1.0×10^{-4}.

If there were a PLL>1 case, one would determine the corresponding required risk reduction factor from the recommended SIL. Then one would take the ratio of this value with the unmitigated frequency as in the first part of the problem to determine the tolerable mitigated event frequency. Then one would use the tolerable mitigated event frequency value, along with the PLL in the intermediate state equation above, to calculate a value for the risk aversion factor α.

10.9a A small release inside the fence line with little damage in table 10.11 is designated C_A. Table 10.8 gives a parameter of W_2 for a 1/10 year frequency. These two values give a "no safety requirements" recommendation in figure 10.2.

10.9b The large release escaping the boundary of the plant with a permanent effect on surrounding environment and likelihood of 1/50 years translates into a C_D, W_1, P_B, and F_B set of categories since there is no probability of avoidance or low occupancy credit that can be taken. This gives row X_6 of column W_1, which indicates an EIL of 3 in figure 10.2.

10.9c The large release escaping the boundary of the plant with a permanent effect on surrounding environment and likelihood of 1/100 years translates into a C_D, W_1, P_B, and F_A set of categories since there is no probability of avoidance credit that can be taken but there is a low occupancy credit since two weeks out of 52 is less than the 10% requirement in table 10.6. This gives row X_5 of column W_1, which indicates an EIL of 2 in figure 10.2.

10.10 Sources of financial loss include:

- Damaged equipment that must be replaced
- Lost revenue due to the inability to produce products (business interruption)
- Penalties imposed by customers for breach-of-supply contracts
- Third-party liability for injuries and fatalities resulting from process accidents
- Fines from regulatory agencies due to personnel injuries and fatalities
- Clean-up costs for environmental damage resulting from spills and releases
- Fines from regulatory agencies due to environmental contamination
- Lost revenue because of decreased sales resulting from public outrage surrounding accident outcomes
- Expense of public relations to repair damaged organizational image

10.11 The overall integrity level 4 should be selected based on the SIL 4 and because the other values are negligible in comparison.

10.12 When multiple types of integrity level are selected for an instrumented function, they should all be listed separately in the safety requirements specification. This process will establish and document the specific requirements for personnel safety, and separate those requirements from the overall integrity level, which may be driven by environmental or financial considerations. The reason for this separation is that most laws, standards, and regulations treat safety, environmental, and financial reasons for instrumented functions differently. If the different criteria are not clearly stated in the specifications, an organization may subject itself to unnecessarily stringent regulation.

10.13 With a 1/50 per year frequency and an expected loss of $55 million, the annual cost of the unmitigated event is $1,100,000. An integrity level 1 system will provide a risk reduction between 10 and 100. In the best case of RRF=100, the frequency will become 1/5,000 per year to give an annual cost of $11,000. This translates into a maximum net annual benefit of $1,089,000, which must be corrected for the cost of the instrumented system itself. In the worst case of RRF=10, the frequency will become 1/500 per year to give an annual cost of $110,000. This translates into a maximum net annual benefit of $990,000, which must also be corrected for the cost of the instrumented system itself.

10.14 The annual cost of the SIF is made up of several components, which are calculated separately and then added together to give the final result. The first piece of data is the installed equipment cost. Depending on the organization's capital cost formula, there could be a range of annualized values for this term. A conservative assumption is to take an annual cost equivalent of 20%, which gives a value of $3,720. The annual testing cost is 4 × $780 = $3,120. The annual maintenance is already given as $350. The annual nuisance trip cost is calculated from the individual trip cost times the trip frequency. The per trip cost is $30,000 + (16/24) × $450,000 = $330,000. Multiplying this by the trip frequency of 1/25 per year gives an annualized trip cost of $330,000 × 1/25 = $13,200. Adding these values together gives the result.

Equipment annual cost	$3,720
Testing annual cost	$3,120
Maintenance annual cost	$350
Nuisance trip annual cost	$13,200
Total annual cost	$20,390

Index